This book is to be returned on or before
the last date stamped below.

AUG. 17. 1982

JAN. 19. 1983

26 JUL 1984

-7. NOV. 1985

28 APR 1987

18 JAN 1993

11 OCT 1994

07 AUG 1995

15th Sept 95

-3 JAN 1996

DUE
1 5 MAR 2011

LIBREX —

PROGRESS IN BIOMASS CONVERSION

VOLUME I

ACADEMIC PRESS RAPID MANUSCRIPT REPRODUCTION

PROGRESS IN BIOMASS CONVERSION
VOLUME I

Edited by
KYOSTI V. SARKANEN
DAVID A. TILLMAN
College of Forest Resources
University of Washington
Seattle, Washington

ACADEMIC PRESS

New York   London   Toronto   Sydney   San Francisco   1979
A Subsidiary of Harcourt Brace Jovanovich, Publishers

Copyright © 1979, by Academic Press, Inc.
ALL RIGHTS RESERVED.
NO PART OF THIS PUBLICATION MAY BE REPRODUCED OR
TRANSMITTED IN ANY FORM OR BY ANY MEANS, ELECTRONIC
OR MECHANICAL, INCLUDING PHOTOCOPY, RECORDING, OR ANY
INFORMATION STORAGE AND RETRIEVAL SYSTEM, WITHOUT
PERMISSION IN WRITING FROM THE PUBLISHER.

ACADEMIC PRESS, INC.
111 Fifth Avenue, New York, New York 10003

United Kingdom Edition published by
ACADEMIC PRESS, INC. (LONDON) LTD.
24/28 Oval Road, London NW1 7DX

ISSN 0192-6551

ISBN 0-12-535901-2

PRINTED IN THE UNITED STATES OF AMERICA

79 80 81 82   9 8 7 6 5 4 3 2 1

Contents

List of Contributors	*vii*
Foreword	*viii*
Preface	*xi*

Living Resources and Renewing Processes: Some Thoughts and Considerations   1
   Ingemar Falkehag

Wood Fuel Use in the Forest Products Industry   27
   R. L. Jamison

The Economic Values of Wood Residues as Fuel   53
   David A. Tillman

Pyrolysis of Wood Residues with a Vertical Bed Reactor   87
   J. A. Knight

Methanol from Wood: A Critical Assessment   117
   R. M. Rowell and A. E. Hokanson

A Survey of United States and European Practices for Recovering Energy from Municipal Waste   145
   James G. Abert and Harvey Alter

The Silvicultural Energy Farm in Perspective   215
   Jean-Francois Henry

Index   *257*

List of Contributors

Numbers in parentheses indicate the pages on which the authors' contributions begin.

James G. Abert (145), Research and Development, National Center for Resource Recovery, Inc., 1211 Connecticut Avenue, N.W. Washington, D.C. 20036

Harvey Alter (145), Research Department, National Center for Resource Recovery, Inc., 1211 Connecticut Avenue, N.W., Washington, D.C. 20036

Ingemar Falkehag (1), Consultant, 706 Creekside Drive, Mt. Pleasant, South Carolina 29464

Jean-Francois Henry (215), Consultant, Route 2, Warrenton, Virginia 22186

A. E. Hokanson (117), formerly with Raphael Katzen Associates, Consulting Engineers, Cincinnati, Ohio

R. L. Jamison (27), Division of Energy Management, Weyerhauser Company, Tacoma, Washington 98401

J. A. Knight (87), Engineering Experiment Station, Georgia Institute of Technology, Atlanta, Georgia 30332

R. M. Rowell (117), Research Specialist, U.S. Department of Agriculture, Forest Service, Forest Products Laboratory, Madison, Wisconsin 53705

David A. Tillman (53), College of Forest Resources, University of Washington, Seattle, Washington 98195

Foreword

THE ROLE OF BIOMASS FUELS

The biomass were perhaps the first fuels used by mankind to meet home heating and cooking needs. In colonial America, a large country with a small and geographically dispersed population, wood was almost everywhere. No matter how remote his location, the settler or homesteader could find wood for fuel nearby. Wood was used to fire the first steamboats and steam locomotives and such early stationary power plants as the Dolbear donkey engine and the Lidgerwood skidder. From the establishment of the Plymouth Colony until the start of World War II, half of all the wood harvested from the forests of the United States was cut for fuel. The per capita consumption of wood fuel peaked at about the time of the Civil War, then declined and only recently rebounded. The easy availability of low-cost fossil fuels, coal first, followed by oil, resulted in their substitution for wood and other sources of biomass as fuel. Forest products manufacturers burned their residues in the form of pulp waste liquor, slabs, edgings, trim, roundup, bark, etc., and thus represented pockets of continued wood fuel utilization. Similarly, sugar refiners continued to generate steam by burning bagasse. A few rural homes continued to depend upon cordwood from the farm wood lot as a source of heat. Tobacco farmers used wood for fuel curing. Many city dwellers used wood in fireplaces more for aesthetic values than for heat, and those addicted to outdoor barbecuing depended upon wood charcoal.

These appeared to be the last bastions of biomass fuel utilization in the United States until the advent of the oil embargo of 1973–1974. Then biomass fuels were rediscovered. The forest products industries, never completely divorced from wood as a fuel, rediscovered its merits as they sought energy independence. Sugar refiners increased their use of bagasse. Cities accelerated their deployment of municipal waste-to-energy plants. The traditional manufacturers of wood stoves suddenly found themselves inundated with orders and expanded production. New manufacturers came into being. Public utilities, manufacturing plants, and public institutions suddenly became interested in wood, crop waste (e.g., corn seed), municipal waste, and manure as either a primary or backup fuel.

The new oil crisis, brought on by political instability in Iran during the winter of 1978–1979, has accelerated interest in biomass for fuel. Despite this renewed interest and the developments just described, biomass still accounts for only about 2.5% of the national energy budget. Is it likely to play a much more important role?

Biomass sources of fuel have many obvious advantages over the fossil fuels. They are renewable materials. For example, a forestry program developed to supply wood fuel could be expected to approximate an even flow of this commodity. Biomass forms are well-known fuels and the basic technology required to make use of them is essentially well known. Certainly improvements in combustion technology are needed, however, if they are to play a more important role in the nation's energy supply system. Again using the wood example, the nation's forests are currently producing no more than one third of their potential output of wood, and even that production is underutilized. Recent studies made at the College of Forest Resources, University of Washington, suggest that wood fuel production in the United States could be in the range of 5–14 quads representing 6–18% of the current national energy budget. This could be substantially increased with more intensive forest management. This use of wood for fuel would not interfere with the production of structural wood products and pulp. By analogy, one can foresee expanded roles for other sources of biomass such as MSW and crop waste.

Biomass sources have some disadvantages as fuels. They are low-density materials, hence bulky to transport. In their natural state, green, they have high-moisture contents that either penalize the heat output of the fuels or require that they be predried.

From an air pollution standpoint, biomass has advantages and disadvantages. When air pollution focuses on health hazards, biomass looks very good. Its SO_2 and NO_x emissions vary from zero to trivial. When air pollution focuses upon particulate matter and visual impact, wood may assume the role of culprit.

Whether biomass will be competitive with fossil fuels and nuclear energy in the market place in the near term depends upon so many intangibles that prediction is difficult, if not impossible. It is clear, however, that it will be competitive in the long term. New increments of fossil fuels come from remote and inhospitable areas of the globe or from deep in the earth or beneath the ocean. The costs in capital, energy, and labor required to add new sources of oil and gas to the supply base are very great indeed. The environmental costs associated with new increments of coal supply are substantial. Biomass supplies on the other hand derive from areas of the earth's surface that are usually quite accessible. Second and successive crops of biomass can reuse the original infrastructure.

Increased use of biomass fuel in substitution for petroleum and natural gas means greater dependence on domestic rather than foreign natural resources. If this is an important objective of a national energy policy, then government may wish to provide some financial incentives to speed up a substitution that is likely to occur in the long run in response to the orderly operation of the market place.

Continued study of the potential of biomass fuel, as represented by the papers in this volume, will provide a technical backup for a biomass fuel utilization effort.

James S. Bethel, Dean
College of Forest Resources
University of Washington
Seattle, Washington

Preface

The high cost in dollars and uncertainties associated with continued over-reliance on petroleum and natural gas as sources of fuel has caused societies throughout the world to investigate many energy alternatives. These alternatives include coal, nuclear power, hydroelectric generation, solar heat, geothermal steam, oil shale, and the biomass. These alternatives have roles to play in the supply of fuels and chemicals now and in succeeding generations. They must be developed to their maximum potential given their unique capabilities and limitations. At this juncture in history, we cannot arbitrarily foreclose on any option or opportunity to replace oil with an appropriate alternative.

Within the context of that many-fuels philosophy, let us consider the biomass sources: wood and wood residues, agricultural materials, and municipal refuse. They are truly unique sources of energy. They are used primarily for high-value products: foodstuffs, fiber, structural materials, and unique chemicals to a lesser extent (e.g., substituted cellulose products). Currently the potential sources of fuel are principally byproducts or residues associated with the production and/or consumption of such high-value products. Sawmill and forest residues, spent pulping liquor, feedlot manure, bagasse, and municipal refuse all fall into this latter category. This intimate association with higher-value commodities at once distinguishes the biomass from other energy sources, since even oil and natural gas are used in limited quantities for nonenergy purposes (e.g., $\sim 3\%$ of total consumption).

The biomass sources exist as natural high polymers in contrast to oil, gas, or coal. The degree of polymerization for cellulose, for example, is $\sim 10,000$. For the hemicelluloses, it is ~ 500. Further, these sources of chemicals can be used without destroying and rebuilding their polymeric structures. This attribute contributes to, causes, their association with higher-value products such as paper, plywood, and food. This attribute places fuels from biomass in a unique position. Fuels must be a relatively inexpensive commodity in an industrial society. Given a strong demand for materials, the best biomass sources are inevitably commanded away from fuels applications in the marketplace.

The other attributes of biomass can be rapidly identified: The sources are available in virtually every region of the nation, they are relatively modest in heat content on an as-available basis (e.g., ~4000 Btu/lb for biomass versus 12,000 Btu/lb for coal and 19,900 Btu/lb for oil), they are virtually sulfur- and nitrogen-free, and they may be low in ash content (~1%). Their low-heat content stems from the hygroscopic nature of biomass fuels leading to moisture contents of 30–50%. Such moisture contents not only dilute the material available for combustion or conversion, but also consume heat as they are vaporized and leave the system in the form of steam.

The implications of these attributes are many. Biomass fuels must be gathered and concentrated rather than mined in concentrated form and distributed to a diverse set of users. Biomass fuels are not readily transportable over distances above ~20–50 mi. Biomass fuels contribute little to environmental degradation, and, in many cases, their use for the production of energy and chemicals solves or alleviates environmental problems.

Biomass fuels, then, are highly useful supplementary sources of energy. They should not be viewed in the same way as petroleum or natural gas, however. Arbitrarily they cannot be forced into the mold of the clean, convenient, concentrated liquid and gaseous fuels that have satiated our energy appetites and have jaded our view concerning what energy sources ought to be like.

There is a great deal of thinking and research currently underway in this area. This research includes techniques for producing higher-quality fuels by pyrolysis, chemicals, and fibers by novel pulping systems, and more. Also emerging are a series of critical assessments concerning such topics as the production of alcohol fuels and the economics of wood fuel use. No consensus exists concerning such topics as fuel farming, biomass liquefaction, or even fuel drying or densification. The biomass energy arena is rife with controversy and contradiction.

It is in the context of this perspective, and the series of controversies now prevalent, that this annual series has been created. Its purpose is to examine aspects of the research and topics of controversy. It is designed to help focus aspects of the debate rather than to forge a consensus. In that respect this series differs from traditional review volumes. Such a divergence is necessary, however, if we are to reflect the current status of activity in the burgeoning biomass energy field.

In the creation of this first volume we are indebted to, and wish to acknowledge the assistance of many individuals: the contributors who provided us with excellent papers; Mrs. Mildred Tillman, who typed all of the manuscripts; the editorial and production people at Academic Press who have been of constant help; and many friends who gave us invaluable advice and aid.

Kyosti V. Sarkanen
David A. Tillman

Progress in Biomass Conversion, Volume I

LIVING RESOURCES AND RENEWING PROCESSES:
SOME THOUGHTS AND CONSIDERATIONS

Ingemar Falkehag

Mt. Pleasant
South Carolina

I.	INTRODUCTION: CARBON RESOURCES AND CYCLES	2
II.	SYSTEMS VIEW AND TIMEFRAME	4
III.	RENEWABILITY OF BIOMASS	6
IV.	PRESENT ORGANIC MATERIALS SYSTEM IN THE U.S.	7
V.	BIOPRODUCTION POTENTIAL	9
VI.	COMPLEMENTARITY OF NATURAL AND INDUSTRIAL SYSTEMS	10
VII.	BIOSYNTHETIC PATHWAYS	15
	A. Photosynthetic Approaches	15
	B. Chemicals from Wood	16
	C. The Chemical Feedstock Approach	17
VIII.	SUBSTITUTION AND SYSTEM STUDIES	17
IX.	INSTITUTIONS	20
X.	IMPLICATIONS FOR SCIENTISTS AND ENGINEERS	20

I. INTRODUCTION: CARBON RESOURCES AND CYCLES

At this time, U.S. consumption of energy, mostly fossil carbon sources, is about equal to the net annual storage of solar energy in the U.S. biomass system. The latter is estimated at about 5 billion tons of biomass per year which in dry form corresponds to a heat value of about 80×10^{15} Btu (Quads). We are indeed at an interesting point in our cultural history. Policies on how we govern the carbon system, including the photosynthesized resources, are indeed pertinent. We are facing some deep philosophical questions concerning how we should manage our organic materials, land, nutrient, and water systems now and in the future. How long can we continue a fossil carbon based industrial development? Will we ultimately have to come back to the solar energy driven carbon system which we were almost totally dependent upon only a century ago?

Will we be wise in shifting back to solar energy and renewable organic resources to meet human needs not only for food, but also for fuels and materials now derived from oil and natural gas? Maybe we will have to make this transition within the next 100 years. In assessing this, it should be kept in mind that most of our problems are systemic in nature. Truly we cannot see the forest for the trees. As Morowitz (1968) has put it, "We are confronting an entropy crisis more than just an energy crisis."

The term "biomass" was introduced by Eugene Adam and relates to the quantity of all living matter from the five kingdoms in biology: plants, animals, fungi, protists, and monerans (Margulis, 1974). Photosynthesis can occur in the protists and monerans as well as in the great phototroph system we call plants, and which will be the resource referred to in this paper. "Bio" is of course derived from "bios" (Greek) meaning life. Living systems are always highly diverse and complex, as are the biomass utilization issues today. They have to be discussed in a holistic context and with a recognition of interdependencies in the overall system of land, water, people, nutrients, and all the five kingdoms of life as illustrated by Table I. The ultimate issues regarding renewable resources will relate to the extent of and wisdom with which humans will demonstrate the other living resources, i.e., how antropocentric we can be. The stock of fossil carbon resources are a great endowment that in an evolutionary perspective has to be used wisely in preparation for a future self-sustaining world.

About 94% of the fossil oil and natural gas resources today are used for fuel purposes. Of the 6% going to the petrochemical industry, only about one third actually is used as material. The energy intensity in production of synthetic organic materials is about 3 tons of oil per ton of product on the average. The competition for some oil fractions, and for natural gas, is likely to intensify and we will see a certain conversion to coal in the petrochemical industry beyond the next 10 years.

It is through utilization of the solar energy flow (subsidy) in the form of stored in fossil carbon reserves that we have been able to carry out the industrial revolution during the last 100 years. This is a short time span in the history of humans and the biosystem, as King Hubbert and others point out (Hubbert, 1971; Hubbert, 1969; Bolin, 1970).

One of the major concerns in the extended transfer of carbon from the fossil sources to the biospheric systems relates to the impact of additional CO_2 generation, half of which is raising the CO_2 level in the atmosphere and half of which is absorbed by the ocean and the biosystem. Some recent studies (Woodwell, 1978; Woodwell, et al., 1978; Stuiver, 1978; Siegenthaler and Oeschger, 1978) indicate that the biosphere might no longer be as effective a sink for CO_2, and that the increase in CO_2 only partly is caused by the burning of fossil carbon. This indicates that the ratio between photosynthesis and biodegradation has decreased. This decrease might be caused either by a reduced rate of the former or an increased rate of the latter. If the stock of

TABLE I. CARBON IN THE BIOSPHERE

Form	Tons $\times 10^{12}$
Carbonate in sediment	18,000
Organic carbon in sediment	6,800
CO_2 in atmosphere	0.65
Living matter on land	0.08
Dead organic matter on land	0.7
CO_2 in ocean	35.4
Living matter in ocean	0.008
Dead organic matter in ocean	2.7

Source: Broda, 1975.

organic matter in soils is actually going down, it could require major changes in land management practices.

The increased absorption of heat radiation by CO_2 should result in a warming trend of the climate. This might be an ultimate concern in relation to how much carbon is handled in the biosystems (National Academy of Sciences, 1975; MIT Press, 1971). However, because of the sun's cyclic activity, we are experiencing a cooling off in the Northern Hemisphere which might be expected to cause droughts and crop failures in the years around 1990. The stock of biomass, mostly forests, can be considered as a food reserve. Policies on future uses of lignocellulosic materials should consider the requirements for adaptation during such discontinuities in the food producing system. The climatic effects of CO_2 in the atmosphere might be of concern only around year 2020 in the Northern Hemisphere, but probably earlier in the Southern Hemisphere.

It appears at this point highly desirable to increase photosynthesis and the gross and net bioproductivity. The management of these processes and the alternative uses of the biomass will be the subject of debate during coming years. The shift in value system from man over nature to man in or with nature plays an important role.

II. SYSTEMS VIEW AND TIME FRAME

In assessing the extended roles of renewable resources, we should not only address the operational or tactical questions of how to alleviate immediate shortages and pressures, but it is also imperative that we act in resonance with strategic and normative considerations. The biomass system assessment by the Office of Technology Assessment can be an extremely important foundation for policy recommendations.

Several raw materials conferences, notably the NSF-AICHE Conference on Chemical Feedstock Alternatives (NSF-AICHE, 1977) and the World Conference on Future Sources of Organic Raw Materials (Toronto Proceedings, 1978), concluded that the most likely scenario for the next 30 years is a gradual shift from oil and gas to coal as a feedstock for the chemical industry (as we now know it). The biomass is projected to play a role beyond year 2000. However, one has to be cautious in generalizing. Government policies, capital availability, and other factors could affect the choices. The Brazilian ethanol project (Toronto Proceedings, 1978) and development work by Gulf Oil Chemicals Company on the uses of low value cellulosics for hydrolysis-fermentation (Emert, 1978) are indicators that in the case of an oxygen containing commodity chemical like ethanol, the time for economic feasibility might come earlier than expected in some of the assessments.

It is clear that natural products will remain competitive as specialty chemicals and where unique macromolecular properties can be used. The emphasis in strategy development can be in determining such niches, properly applied to regional and national conditions (Houston Proceedings, 1978).

It is apparent that the further assessment will require an interdisciplinary effort and a general systems approach with consideration for hierarchial levels (Miller, 1971), the complexity and desirable diversity and adaptability of natural systems, the cyclic nature of materials and energy flow patterns in renewable systems, etc. The vertical and horizontal integration we talk about in industry is used perpetually in nature to improve survivability.

Two questions immediately come up in considering renewable resources for new and extended uses as feedstocks:

(1) Is it technically feasible to produce the major petrochemicals and polymers from renewable resources?
(2) Is there, in the U.S., enough renewable resources available for a shift from oil as a raw material without adversely affecting food, lumber and paper production?

The answers to these questions today appear to be yes. However the substitution for oil and gas in polymer and organic materials production is not so much a matter of technical feasibility and resource availability as a matter of driving forces, constraints and uncertainties affecting a change. The energetics in producing a product from alternative raw materials vary and can be in favor of renewable resources. Optimum plant size, logistics, labor intensity and the cost and availability of capital enter into the economic picture. The environmental and social costs in relation to alternatives have to be assessed. Traditional economics does not account for renewability or nonrenewability. It makes no distinction between reversible and irreversible processes. Georgescu-Roegeu (1971) has discussed the need to account for the entropic loss with nonrenewable resources. The concept of economics of scale is being challenged by Schumacher (1973) and others and terms such as appropriate technology are increasingly heard of. Some of these emerging concepts are more applicable to renewable resources than to fossil carbon sources. The competitiveness of natural rubber compared to the synthetic products is a case in point. The assessment of renewable resources uses thus has to include not only aspects of what we call economics, but also environmental, social, and political factors. As Sarkanen (1977) has pointed out, "the area should be looked at as a whole, rather than having separate groups of parochial researchers concentrate on forest residues, waste products from the pulping industry, agricultural residues, or marine resources. This calls for a broader interdisciplinary endeavor than is possible in the framework of existing government agencies."

One can amplify and extend that concept and add a warning about the simplistic, plug-in approach of producing petrochemicals from wood. It is likely that we will continue to see integrated systems, similar to the present lumber-board-paper-tall oil-energy system. We should stay at highest possible systemic levels. The energy farm as a single output system is only justifiable if markets and needs for higher value materials than energy do not exist. An immediate issue is how we can upgrade renewable resources that today are wasted or used for energy production.

III. RENEWABILITY OF BIOMASS

The Board on Agriculture and Renewable Resources of the National Academy of Sciences organized the CORRIM program. CORRIM defines as "renewable" a material that can be restored when the initial stock has been exhausted. The dynamic nature of the concept of renewability is recognized. A renewability ratio is defined as the ratio of replenishment rate to depletion rate. Broadly speaking, renewable resource is used as a synonym for a resource of biological origin while nonrenewable resource is used as a synonym for a resource of geological origin.

A carbon atom in a biological material might have its origin in oil or coal or even a mineral such as calcium carbonate. In the case of these fossil fuels, however, the carbon did originate in biological systems and was synthesized by nature into the more concentrated fuel forms. The energy source that causes the renewing is the sun for the phototrophs (autotrophs), the plants and the photosynthesizing bacteria. Electromagnetic radiation and gravitational forces give the energy flow in biology that has driven evolution and produced our biomass stock and fossil carbon sources. The enormous bioproductivity of the salt water marsh (Spartina alterniflora) is possible because of solar radiation and tidal pulsation. We have in that case a sun and moon powered system. The water splitting by light quanta resulting in CO_2 reduction starts the process. In fact, our primary concern should be with the process of renewing our resources.

We have a classical matter-energy and structure-process issue. Renewable resources can be looked upon as a temporarily frozen solar energy process.

IV. PRESENT ORGANIC MATERIALS SYSTEM IN THE U.S.

The use intensity of new supply of materials has been discussed by Radcliffe (1973). The per capita consumption of synthetic polymers (derived from fossil sources) constitute only 6% of the total organic materials consumption, and thus renewable materials today are consumed at a rate 16 times greater than nonrenewable organic materials. This ratio is shown clearly in Table II, comparing synthetic polymers to total renewable resources. This is shown in more detail in Table III.

The increase in rate of materials and energy consumption follow each other closely. As pointed out by Keyfitz (1976), the growth attributable to affluence is greater than the population growth.

TABLE II. *Use Intensity of New Supply of Materials in the U.S.*

	Lbs per capita for 1974
Nonrenewable resources	
Nonmetallic minerals	18,900
Metals	1,340
Synthetic polymers	148
Total	20,388
Renewable resources	
Wood and wood products (1971)	2,222
Fibers (other than wood)	29
Natural rubber	8
Leather	14
Total	2,273

Source: Radcliffe, 1973.

TABLE III. Organic Materials Production and Use in the U.S. (Approximate figures, 1972-1974)

	Tons × 10^6/year				
	Food-feed	Materials	Energy	Residuals	Total
Synthetic polymers	–	18	(36)	–	
Lumber and rigid panels	–	119	16	25	160
Paper and paperboard	–	57	30	–	87
Forest residues	–	–	–	150	
Noncommercial timber	–	–	–	100	
Municipal waste	–	–	–	100	
Bushes, shrubs, foliage	–	–	–	>50	
Hardwoods on pine sites	–	–	–	>50	
Cotton	–	3	–	13	
Fats and oils	6	3	–	–	
Soybeans and peanuts	40	–	–	65	
Grain crops	250	2	–	300	
Forage	240	–	–	–	
Sugar crops	10	0.2	5	6	
Animal wastes	–	–	–	360	
Approximate totals	550	200	50	1200	2,000
Total net biomass production					5,000

Published data[1] on the production of renewable organic resources and various uses and nonuses vary considerably but an attempt has been made in Table III to differentiate between food-feed, materials, energy, and residuals or unused material. The latter group are generally referred to as "waste." Some

[1] See the following references: Abelson, 1973; NAS, 1976; Forest Service; Wilke, 1975; Ctr., Metrop. Stud., 1976; Applied Polymer Symp., 1975; Barr, 1976; Proceedings-Toronto, 1978; Int. Conference-Houston, 1978.

figures are estimated and several resources have not been listed. The noncommercial timber stock is estimated to over 1 billion tons but this can, of course, not be the annual out-take. The recoverable quantity of the residuals depends on economic and environmental considerations.

It seems likely that the consumption of renewable resources for the traditional materials (lumber, plywood, particle board, flakeboard, fiber board, insulating board, paper, paperboard and hardboards) will at least double by the year 2000 (Radcliffe, 1973). The primary needs of the forests products industries relate to reduction in energy intensity, better raw material utilization, less capital intensive processes and improve environmental control processes. The CORRIM study (NAS, 1976) discusses future needs in the conventional structural and fiber uses of wood.

V. BIOPRODUCTION POTENTIAL

Human activities in the U.S. interfere with about 25% of the net biomass production through various forms of harvesting, but probably only about 15% leaves the land. Some of this used biomass (food, feed and materials) is again returned to the soil.

Various forms of management techniques such as fertilization, pest control, irrigation, genetic plant selection, and thinning can improve productivity considerably and the recommendation for increased productivity made for agriculture (NAS, 1975) can in principle also be applied to forestry and biomass plantations.

The 500 million acres of commercial forest land has a net annual productivity of less than 1 ton per acre. The biological potential (Radcliffe, 1973) of 400-500 million tons per year can probably be increased by at least 50%. Whole tree utilization concepts (Applied Polymer Symp., 1976) are being adopted and intensive, short rotation forestry of hardwoods can give yields of up to 4 tons per acre a year. A primary concern in the use of these intensive techniques relates to the tolerable removal of organics and nutrients from the soil (Boyle, 1976) and other environmental impacts (Davis, 1976; Applied Polymer Symp., 1976).

In addition to the commercial forest land, there are 250 million acres of noncommercial forests of which 20 million acres are assigned as parks, wilderness areas, and other deliberate land withdrawals. The forests totally occupy about 1/3 of the U.S. land area. The use of nonforest, nonagricultural land for biomass production should be the subject

Intensive biomass production on land or water can give yields of up to 30-50 tons per acre a year for C_4 plants under optimum conditions (SRI, 1974; Center for Metrop. Stud., 1976). Thus it appears that production of lignocellulosic materials can remain complementary with food production and that, depending on population growth rate and international developments, adequate quantities of nonfood biomass will be available for materials, including synthetic polymers, if necessary. In fact, it is likely that in some regions of the world the distinction between agriculture and forestry will be less pronounced, as indicated by the emergence of concepts such as agroforestry and treecrops.

With a timeframe of more than 30 years and with continuation of present growth rate increases, major stresses will more than likely occur in the organic materials and land use systems. It seems plausible that new patterns of materials use will have to develop before that time. It is now appropriate to see how we can harmonize our use patterns with the production capacity of the photosynthetic system (Proc.-Toronto, 1978; Int. Conf.-Houston, 1978). The multiple interdependencies makes this a very complex task.

VI. COMPLEMENTARITY OF NATURAL AND INDUSTRIAL SYSTEMS

What is nature then capable of producing qualitatively, and how can the biosynthesized materials meet shifting human needs? Have our materials requirement in terms of performance, as achieved through the marvelous developments in polymers and composite materials, deviated so much from the properties-performances of natural materials that we have to increasingly rely on the "feedstock approach" of using renewable resources as another carbon raw material source, comparable with coal, shale oil, lignite and peat?

If we separate between bioproduction, conversion, and consumption, we can look at the capability of the solar energy driven production system to produce molecules and structures at various free energy levels which have to be modified to meet the thermodynamic requirements of the human consumption system (Center for Metrop. Stud., 1976). The symbiotic relationship between the earth and humankind has been discussed by Dubos (1976).

The hierarchial levels of the natural materials system is shown in Table IV. Only a single example is given at each level. The manner in which we go down the systems scale is of course a primary question. The cost of going down the scale to meet a social need can be expressed in energetic

TABLE IV. Hierarchial Material System Levels

Biosphere

Autotrophs	*Heterotrophs*
Forest	*Human society*
Tree	*Supply system*
Stem	*Transportation*
Wood ⟶	*Pallet and goods*
Fiber ⟶	*Box*
Cellulose fiber and fibril	*Carton (paper)*
Microfibril and protofibril	*Film barrier*
Cellulose molecule	*Polyethylene*
Glucose ⟶ *Ethanol* ⟵	*Ethylene*

---

terms. In principle it should be desirable to cross over from the production system to the consumption (human) system at highest possible level, e.g., wood for pallets and wood fibers for paper.

The lignocellulosic plants which by far constitute the greatest stock of biomass on earth (2.10^{12} tons), are not digestible by man, but can be made digestible for ruminants. The foliage is, however, directly digestible by various animals and could be a source of protein for man if adequate collection and separation processes were developed (Applied Polymer Symp., 1976). The foliage can constitute up to 7% of the weight of the plant and can contain up to 8% of protein for hardwoods (half as much as alfalfa). The sugar and the protein (legumes) or starchy type plants have generally more than 50% lignocellulosic material in the roots, stem and branches. Often it is the seeds which we eat.

From a chemical point of view, we can group the materials into carbohydrates, phenolics, proteins, lipids, and special biomolecules, such as chlorophyll, vitamins, etc. The component roles can be as building stones and adhesives, energy sources, synthesizers, environmental protectors (stress adjusters), participating in both anabolic and catabolic processes (Miller, 1971).

We hear often about cellulose as being the major polymer on earth. In terms of volume and weight, this is true, but in terms of storage of solar energy, lignin is the dominating biomaterial. Trees have 35-45% cellulose and 20-30% lignin, but lignin has almost twice as high enthalpic level (heat of combustion) as cellulose as Table V shows. Nature has presumably a purpose in this.

Some of the functions of lignin can be described as (Falkehag, 1975):

Response to stresses: Mechanical

Biochemical (degradation)

Physical-chemical (water)

Chemical (O_2, O_3, UV-light, fire)

Energy storage

Contributions to soil (humus) properties

Cellulose and hemicelluloses might have relatively simple composite materials functions in wood, while the protein and chlorophyll have very complex functions. The lipids might act as surfactants, hydrophobizing (sizing) agents and agents for control of insects, fungi and diseases. A better understanding of the functional roles of plant components and means of affecting their biosynthesis should have high priority as a research area. We know a considerable amount about the organic chemistry of plant components, but much less about the biosynthesis and the manner in which the molecular,

TABLE V. Heats of Combustion for Some Plant Components

Compound	ΔHc (25°C)	
	Btu/lb	Cal/gm
α - pinene	19,600	10,900
Oleic acid	17,000	9,500
Lignin	12,700	7,100
Cellulose	7,500	4,200
CO_2	0	0

Compound + O_2 $\xrightarrow{\Delta Hc}$ CO_2 + H_2O

(State of Reduction ↑)

macromolecular and morphological structural features relates to processes and property-performance-requirement characteristics of the plant.

The free energy in various plant components relates to functions. In the human system we can simply use the enthalpic value and burn the biomass for energy or we can attempt to use renewable resources at highest possible systems level (see Table IV). We should not increase entropy and destroy a composite material, a fiber or a macromolecule when we do not have to (to meet our need). The manner in which we manipulate the biomass and make: (1) cross levels transformations and (2) changes at the molecular level by changing carbon-hydrogen-oxygen balances, can be the framework for important research policy recommendations. To a certain extent, these questions can be approached through thermodynamic tools (Morowitz, 1968; Lehninger, 1973). Work on natural products in this area is badly lacking as the petrochemical interests have controlled thermodynamcis research.

From material science point of view, the research field is open. We do not even know much about the composite materials contributions of the various components in wood. The interplay of natural products at various systems levels with synthetic polymers and inorganic materials has room for many innovations. A definition of materials performance requirements is often the bottleneck. Table VI shows some material system types, many of which already are used for natural products.

TABLE VI. Materials Systems

Type	Examples
Uniform, amorphous	Lignin-phenol resin
Partially chrystalline	Rayon (cellulose)
Laminated sheets	Plywood
Fiber network	Paper
Bonding agent	Rosin adhesive
Fiber reinforcement	Wood
Particle reinforcement	Lignin-rubber composite
Polyblend	Wood middle lamella
Coating	Starch

A better understanding of structure-process relationships at various hierarchial levels (Baer, 1974) is much needed. In fact, general systems science could contribute considerably to the renewable materials understanding.

Below is given one example of where a renewable resource (lignin) can substitute for a nonrenewable material (carbon black). Adequate (but slightly different) properties in the reinforcement of rubber can be achieved with a lignin replacing HAF or ISAF carbon black, as is illustrated in Table VII. Lignin as a reinforcing filler (below 1000 Å) is not like carbon black in its properties, and hypotheses on filler parameters' effect on materials properties cannot be extrapolated (Falkehag, 1975).

The abrasion resistance with a lignin reinforced rubber does not appear to be governed by the failure properties but rather by the visco-elastic properties of the cured rubber. Lignin is a macromolecular material with lower modulus and hardness than carbon black. The modulus of the reinforcing particle has been shown to affect the reinforcement properties.

The shift from carbon black to lignin in the rubber industry is primarily controlled by institutional factors, lack of economic incentive and concern for pulp mill impacts by recovery of a large fraction of the lignin which has to be replaced with another fuel source with present recovery systems. The quantity of lignin burned annually in U.S. kraft pulp mills is about 16 million tons.

TABLE VII. Physical Properties of Lignin and Carbon Black Reinforced Styrene-Butadiene Rubber at 68 Parts Lignin per 100 Parts Rubber (Oil-Extended SBR)

	Lignin A	Lignin B	HAF	ISAF
Modulus (psi)	520	650	610	730
Tensile strength (psi)	3165	3380	2500	2930
Elongation (%)	720	630	720	750
Tear resistance (ppi)	355	300	320	335
Hardness (Shore A)	54	54	56	61
Corrected pico abrasion	86	85	91	114

VII. BIOSYNTHETIC PATHWAYS

Before discussing the feedstock approach of producing chemicals from renewable resources, it might be useful to look at the photochemistry and biosynthetic pathways of making chemicals, an area emphasized by Calvin (1975) for many years. Solar energy can be used for both heat and quantum collection. In the latter category falls photosynthesis, photochemistry and photoelectric processes.

A. *Photosynthetic Approaches*

The primary and most important step in photosynthesis does not have to do with carbon but is the split of H_2O leading to oxygen and highly reduced products which can affect the CO_2-reduction. The carbohydrate synthesizing cycles are then the starting point for synthesis of proteins, lipids, and phenolics.

A conscious human effort to design photosynthetic systems (plants, bacteria or nonlive systems) to produce food, materials and energy for internal as well as external metabolic systems might be as important an evolutionary event as the domestication of plants and animals in what we call agriculture. Philosophical questions of maintaining (increasing) diversity and complexity to safeguard adaptability get into the picture in considering the further domestication of biosynthetic pathways.

Practical examples of controlling the production of specific chemicals are the natural rubber and naval stores industries. Termite resistant, resin loaded pine beams were once produced in the South. Ongoing efforts to triple the production of rosin and turpentine by chemical stressing of pine is actively studied by Forest Service (Lightwood Research Coord. Council, 1976). Ecological impact is of major concern in this project and the bioenergetics in relation to endproduct value has to be researched.

Zaborsky (Center for Metrop. Stud., 1976) has proposed a long range strategy of bioconversion using regulated plants or microbes or isolated cellular components for the selective production of small active molecules. The argument would be that photosynthesized macromolecules and plant components cannot be made to meet material needs and that fragmentation processes are expensive, energy consuming and requiring complex separation processes as multiple (watersoluble) products are formed. An exception of easy separation is methane from anaerobic digestion. If we need other hydrocarbons, we can, as Ehrensvärd (1975) has proposed, achieve an enzymatic instant fossilization, but this would be quite expensive.

The photosynthetic system can be used to produce chemicals by:

(1) Modifying productivity of existing organisms.
(2) Affecting the selective component synthesis with existing organisms.
(3) Interference with biosynthetic pathways (e.g., catch intermediates).
(4) Biosynthetic production of complex molecules with needed properties.
(5) Photosynthetic feed stock approach by production of small molecules such as H_2, O_2, H_2O_2, CH_4, CH_3OH, CH_2O, CO, NH_3, C_2H_2, and C_2H_5OH.

B. Chemicals from Wood

The direct recovery of chemicals from wood has recently been reviewed in an excellent manner by Hergert and Herrick (1977); Barton, (1978);and Goldstein (1978). The primary economic uses for cellulose and lignin today rely on the macromolecular properties and involve various types of chemical modification giving the added value of the products as specialty chemicals. The same is true for the lower molecular weight constituents, particularly the tall oil components which have become increasingly competitive with synthetic products. Many of the applications for wood chemicals today use some surface chemistry function which generally means a relatively limited market volume. The commodity in large volume applications are in the polymer-materials field where natural polymers have had difficulties with competing synthetic polymers because of economics, reproducibility and quality assurance.

The incentives for uses of natural polymers are, however, likely to increase in coming decades and the research and development should now be increased, particularly as related to composite materials applications. The general philosophy should be to identify and enhance the unique properties of natural macromolecules and to recognize the limitations because of chemistry or molecular configuration. An understanding of the material property contributions of cellulose, hemicelluloses and lignin in the wood can aid in the identification of most suitable commercial applications. The high tensile strength properties of cellulose and the adhesives and reinforcing properties of lignin should be used in the most appropriate composite materials combinations.

Very likely future applications will involve a marriage with synthetic polymers and inorganic materials. The time now appears to be appropriate for a renewed research effort on natural polymers. An important role can be visualized for the polymer and materials departments at universities which have almost totally concentrated on synthetic polymers derived from petrochemicals during recent decades.

C. *The Chemical Feedstock Approach*

Various assessments[1] indicate that abundant biomass resources are potentially available for chemical conversion and that conversion of lignocellulosic material to glucose, ethanol, syngas, methanol, furfural and phenol are technically possible, although in most cases demonstration work is required and optimization has to be achieved. The economics at present energy and wood cost does not yet appear to justify production of bulk chemicals from wood or waste, but considerable uncertainties still exist on actual costs. If renewable resource derived chemicals or substitutes are less energy intensive than fossil carbon derived chemicals, a substitution might be justified.

Uncertainties about coal conversion processes adds to the difficulties in decision making. More information about differences in conversion costs, labor and social costs are needed and justify extensive federal funding for research, development and demonstration project. The two major types of feedstock chemicals are the olefins and the aromatics. Carbohydrates are most conducive for conversion to the former while lignin can be a source of aromatics. Coal will likely be a more economic source for aromatics than for olefins.

VIII. SUBSTITUTION AND SYSTEM STUDIES

As pointed out earlier, all of these assessments have to consider multiple interactions in the energy-materials system. The concepts of net energy and energetics in materials production can usefully be applied. Berry (1973) has discussed the thermodynamics and energetics of alternative materials in various processes. Hoffman, *et al.* at Brookhaven National

---

[1]*See the following references: Radcliffe, 1973; NAS, 1976; USDA-Forest Service; Pulp and Paper Res. Inst., 1975; Goldstein, 1978; Proceedings-Toronto, 1978; Proceedings-Houston, 1977.*

Laboratory (Proceedings-Toronto, 1978) have developed guidelines for a Reference Materials System similar to the Energy Reference System. The general outline of this system is shown in Figure 1. This can be an extremely useful tool developing a framework for a materials policy. Hoffman's input to the systems group in the CORRIM study (NAS, 1976) has led to a preliminary trajectory for the renewable resource system. It provides a quantitative materials flow and some inputs of the energy requirement at the various steps from growing and harvesting to final use.

The technique can also include inputs of labor, capital and environmental activities, and might be particularly useful in studying the effects of perturbations in the various parameters.

In analyzing the appropriate circumstances for a substitution of a feedstock chemical or a final product or need satisfying system, it is important to describe the underlying value system. Traditionally this has been based primarily on economic criteria. However, if a high value will be identified for a reduction in all imports, new forces affecting substitutions can come into play. Below are given some bases for judgment of the desirability of a substitution. They can be seen in a hierarchial and maybe evolutionary perspective, and the criteria are not necessarily competitive.

(1) Traditional economics is based on demand and supply equilibria in its pure form. Decisions are based mostly on the past experiences plus some form of trend extrapolation;

(2) Regulated economics with national or regulated goal incorporated, e.g., reduction in all import or job creation;

(3) Thermodynamic criteria, considering both quantitative and qualitative aspects and focusing on what is right from an energetic point of view. Net energy analysis is a tool in assessments;

(4) Foresight economics with a futures accounting, considering the entropic loss (Georgescu-Roegeu, 1971) of depletable resources and the value of renewability of (photo-synthesized) materials; and

(5) Ecological and moral value judgments related to the "carrying capacity" of the biosphere and the wellbeing of its organisms, presumably with primary emphasis on humans but with a recognition of the value of living with nature without overexploitation. The flow economics of Simmonds (Int. Conf. Houston, 1978) in the context of an ecology of change is particularly well applicable to the bio resource development situation.

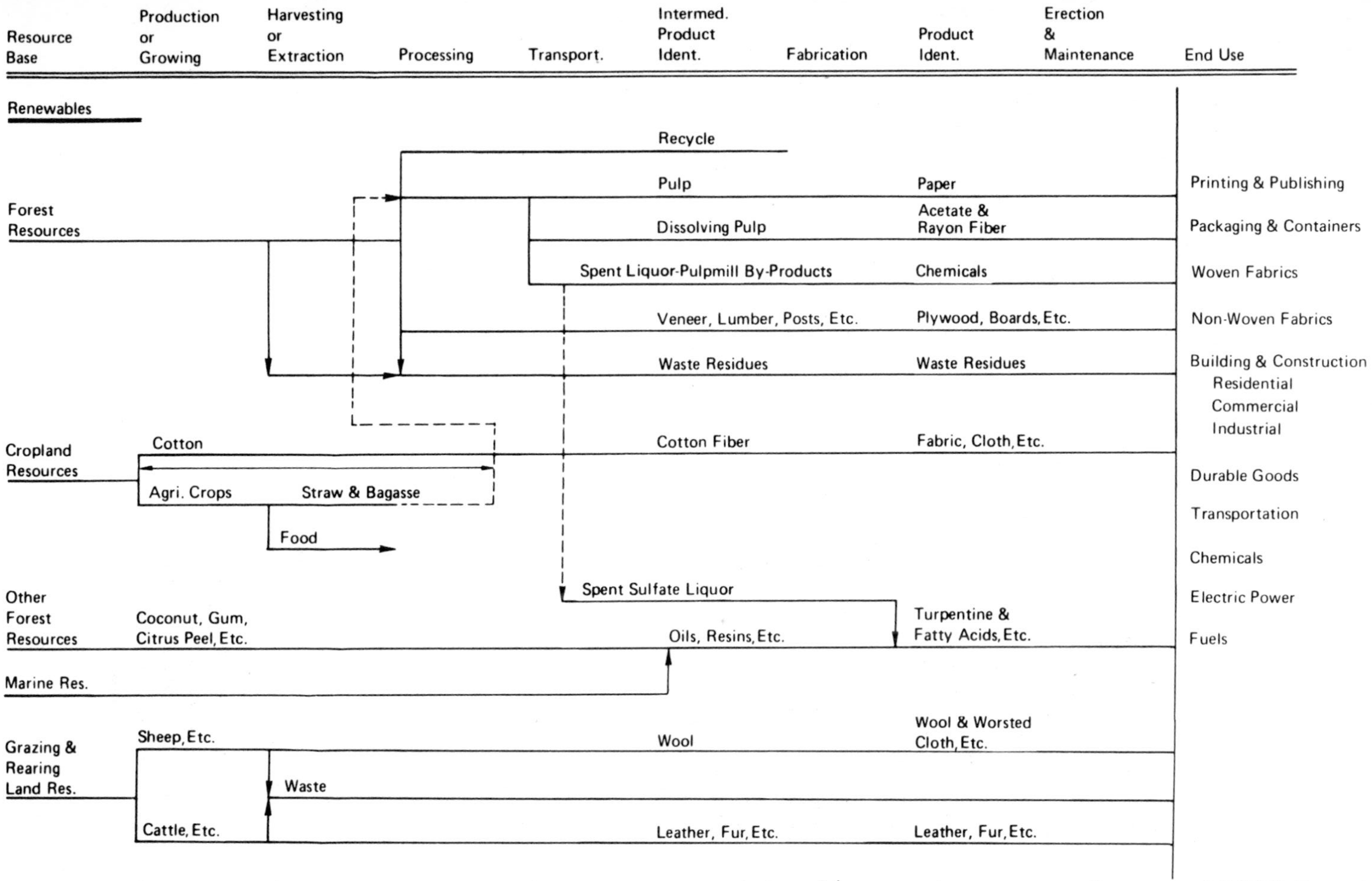

FIGURE 1. The generalized reference materials system applied to biomass resources. Source: CORRIM Report, 1976.

IX. INSTITUTIONS

Just as we are emphasizing renewable resources rather than processes of renewal and the energy-materials flows, it seems that institutions are mostly looked upon as structures with well determined processes. This might be true as long as we have a homeostatic system with agreed upon ends and purposes. However, a consensus is emerging that the research and educational system in relation to renewable resources and the forest industry should be revitalized.

The emphasis on structure shows up in the names of institutes and departments (wood, cellulose, forest products, and paper) although a recent trend has been to include environmental science and thus obtain a more interdisciplinary outlook.

If we are approaching a state where we will view our photosynthesized resources in a new way with regard to generation, conversions and end-uses, we might not be able to rely on trend extrapolation. As scientists we might, in fact, confront a major paradigm shift (Kuhn). Harman (1975) at Stanford Research Institute has compared a transformation perspective with the Kahn post-industrial perspective. Henderson (1975) and Beer (1975) and others have applied the metasystem thinking about institutional change and concepts of managed rather than exploited resources. Emerging understanding of self-organizing systems (Glandsdorff, 1974) and the evolutionary view of Jantsch (1975) could be of particular relevance in dealing with questions of our interdependencies with the photosynthesizing systems and the resulting products: food, renewable resources and other bioproducts.

In summary, we might talk about renewable resources through renewable organizations and institutions.

X. IMPLICATIONS FOR SCIENTISTS AND ENGINEERS

The renewable resource and materials renewal issue are involving major uncertainties and high complexity (multiple interdependencies) with regard to extended and new uses. The timeframe for change is important to consider in the nonconventional uses of renewable resources. Considerations about the total biomass system and the mutualities between forestry and agriculture adds to the need for interdisciplinary and systems oriented views of the pattern of change.

The existing areas of renewable resource use confronts such needs as:

(1) safeguarding raw material supply,
(2) less capital intensive technology,
(3) less energy intensive processes,
(4) improved environmental control,
(5) less dependency on depletable resources, and
(6) better utilization of all resources.

The extended or new uses of renewable resources raises challenges in many areas in relation to the production, conversion and uses of renewable resources. Most traditional institutions are not very well oriented towards handling some of the tasks ahead, and this is particularly true in the material science and engineering areas.

The age of substitutability has been used as a description of our present materials situation. The implications of a transformation in the renewable resource system should be the subject of well organized assessments, incorporating technological, economic, social, environmental, and educational concerns.

Below are given some examples of research, development, and demonstration activities at three planning levels:

A. *Bioproduction*

(1) *Normative level*. Develop an awareness about the functional roles of the components in phototrophs, the ability to direct the selective production of valuable components and the manner in which plant components, macromolecules and chemicals can best be integrated (symbiotically) with the human needs system, using the biosynthesized product at highest possible systems level.

(2) *Strategic level*. Develop joint forestry-agriculture programs in such areas as biological nitrogen fixation, water management, genetic selection of plants (for optimum production of a combination of plant components for food, materials, chemicals and fuel), nutrient flows and tolerable biomass removal from ecosystem, etc.

(3) *Operational level*. Survey the existing biomass systems with regard to type, quantity of different plants, economics of harvesting and transportation to potential use sites.

B. *Harvesting, collection, transportation, processing, conversion and fabrication to needed products*

   (1) *Normative level*. Assess alternative socioeconomic systems for ecologically acceptable transformations of photosynthesized materials to end products meeting human needs in an adaptable manner (according to shifting priorities).

   (2) *Strategic level*. Develop improved methodology of substitution analysis enabling the assessment of the benefits and constraints in choosing alternative raw material sources for specific functional and products.

   (3) *Operational level*. Demonstrate technical feasibility and economics of integrated production of ethanol, furfural and phenol from wood.

C. *Product development and use*

   (1) *Normative level*. Determine the structure-property-performance relationships for materials components and systems derivable from renewable resources, assess future organic materials requirements and substitutions and develop approaches for optimization (according to "ortho philosophy") of the use of renewable resources to meet materials needs in manners compatible with food and other needs.

   (2) *Strategic level*. Develop relevant composites and polyblends using renewable resource materials and macromolecules in combination with synthetic polymers (when necessary for performance) and inorganic materials.

   (3) *Operational level*. Assess the feasibility of using modified lignins as adhesives for reconstituted wood products.

REFERENCES

Abelson, P. H. and Hammond, A. L., eds. (1973). "Materials-Renewable and Nonrenewable Resources." AAAS, Washington, DC. (Also see *Science 191* (4228).

Applied Polymer Symposia 28 (1975). "Wood Chemicals—A Future Challenge," Proceedings of the Eighth Cellulose Conference. Wiley-Interscience, New York.

Applied Polymer Symposia 28 (1976). "Complete Tree Utilization and Biosynthesis and Structure of Cellulose," Proceedings of the Ninth Cellulose Conference. Wiley-Interscience, New York.

Baer, E., Gathercole, L. J., and Keller, A. (1974). Structure Hierarchies in Tendon Collagen: An Interim Summary," Proceedings of 1974 Colston Conference.

Barr, W. J. and Parker, F. A. (1976). "The Introduction of Methanol as a New Fuel into the United States Economy." American Energy Research Company, March.

Barton, G. M. (1978). "Chemicals from Trees, Outlook for the Future." Paper at the Eighth World Forestry Congress in Jakarta, October 16-28.

Beer, S. (1975). "Platform for Change." Wiley-Interscience, New York.

Berry, R. S. and Makino, H. (1973). "Consumer Goods—A Thermodynamic Analysis of Packaging, Transport, and Storage," Chicago Institute for Environmental Quality, Chicago, Illinois.

Bolin, B. (1970). The Carbon Cycle, *Sci. Am. 223*, p. 124.

Boyle, J. R. (1976). A System for Evaluating Potential Impacts of Whole Tree Utilization on Site Quality, *TAPPI 59*, 79.

Broda, E. (1975). "The Evolution of the Bioenergetic Processes." Pergamon.

Calvin, M. (1975). "Photosynthesis as a Resource for Energy and Materials." National Science Foundation, Washington, DC.

Center for Metropolitan Studies (1976). Proceedings from Conference on Capturing the Sun through Bioconversion, Washington, DC, March 10-12.

Davis, R. F. (1976). The Effect of Whole Tree Utilization on the Forest Environment, *TAPPI 59*, 76.

Dubos, R. (1976). Symbiosis Between the Earth and Humankind, *Science 193*, 459.

Ehrensvärd, G. (1975). "Fuel Production from Wood," Wood Chemicals Symposium, Syracuse, New York, May 19-23.

Emert, G. H. (1978). "Chemicals and Fuels from Papermill Sludges." Paper presented at the 1978 NCASI Southern Regional Meeting, Atlanta, Georgia, August 2-3.

Falkehag, S. I. (1975). "Lignin in Materials," Applied Polymer Symposium No. 28, 247, Wiley-Interscience, New York.

Georgescu-Roegeu, N. (1971). "The Entropy Law and Economics," Harvard University Press, Cambridge, Massachusetts.

Glansdorff, P. and Prigogine, I. (1974). "Thermodynamic Theory of Structure, Stability and Fluctuations," Wiley-Interscience, New York.

Harman, W. W. (1975). Notes on the Coming Transformation, *in* "The Next 24 Years—Crisis and Opportunity," World Future Society.

Henderson, H. (1975). "No! to Cartesian Logic!" World Future Society Meeting, June 2-5.

Herrick, F. W. and Hergert, H. L. (1977). Utilization of Chemicals from Wood: Retrospect and Prospect, *in* "Recent Advances in Phytochemistry," Vol. II, pp. 443-515.

Hubbert, M. K. (1969). Energy Resources, *in* "Resources and Man," p. 157, Freeman, San Francisco.

Hubbert, M. K. (1971). The Energy Resources of the Earth, *Sci. Am. 224*, 60.

Goldstein, I. S. (1978). "Chemicals from Wood: Outlook for the Future." Paper at the Eighth World Forestry Congress in Jakarta, October 16-28.

International Conference on Bio Resources for Development (1978), Houston, Texas, November 6-10.

Jantsch, E. (1975). "Design for Evolution," Braziller.

Keyfitz, N. (1976). World Resources and the World Middle Class, *Sci. Am. 235,* 28.

Kuhn, T. S. "The Structure of Scientific Revolutions," International Encyclopedia of United Science, University of Chicago Press, Chicago, Illinois

Lehninger, A. L. (1973). "Bioenergetics," Benjamin.

Lightwood Research Coordinating Council (1976). Proceedings of Annual Meeting, January 20-21.

Margulls, L. (1974). The Classification and Evolution of Prokaryotes and Eukaryotes, *in* Handbook of Genetics, Vol. 1.

Miller, J. G. (1971). The Nature of Living Systems, *Behavioral Science 16,* (4) 277

Morowitz, H. J. (1968). "Energy Flow in Biology." Academic Press, New York.

National Academy of Sciences (1975). "Understanding Climatic Change—A Program for Action," Washington, DC.

National Academy of Sciences (1975). "World Food and Nutrition Study—Enhancement of Food Production in the U.S., Washington, DC.

National Academy of Sciences (1976). Committee on Renewable Resources for Industrial Materials, Report, July, Washington, DC.

Proceedings of the Conference on Chemical Feedstock Alternatives (1977). Sponsored by NSF and AICHE, Houston, Texas, October 2-5.

Proceedings of the World Conference on Future Sources of Organic Raw Materials (1978), Toronto, July 10-13. Pergamon Press.

Pulp and Paper Research Institute of Canada (1975). "Feasibility Study of Chemical Feedstock from Wood Waste."

Radcliffe, S. V. (1976). World Changes and Chances: Some New Perspectives for Materials, *Science 191* (4227).

Sarkanen, K. V. (1976). Renewable Resources for the Production of Fuels and Chemicals, *Science 191* (4227).

Schumacher, E. F. (1973). "Small is Beautiful." Harper and Row, New York.

Siegenthaler, H. and Oeschger, H. (1978). Predicting Future Atmospheric Carbon Dioxide Levels, *Science 199*, 388-394.

Simonds, P. (1978). International Conference on Bio Resources for Development, Houston, Texas, November 6-10.

Stanford Research Institute (1974). "Effective Utilization of Solar Energy to Produce Clean Fuel." Palo Alto, California.

The Study of Man's Impact on Climate (1971). "Inadvertent Climate Modification." MIT Press, Cambridge, Massachusetts.

Stuiver, M. (1978). Atmospheric Carbon Dioxide and Carbon Reservoir Changes, *Science 199*, 253-258.

USDA-Forest Service, "The Feasibility of Utilizing Forest Residues for Energy and Chemicals." Washington, D.C.

Wilke, C. R., ed. (1975). "Cellulose as a Chemical and Energy Resource." Wiley-Interscience, New York.

Woodwell, G. M. (1978). The Carbon Dioxide Question, *Sci. Am. 238* (1), 34-43.

Woodwell, G. M., Whittaker, R. H., Reiners, W. A., Likens, G. E., Delwiche, C. C., and Botkin, D. B. (1978). The Biota and the World Carbon Budget, *Science 199*, 141-146.

Zaborsky, O. (1978). International Conference on Bio Resources for Development, Houston, Texas, November 6-10.

WOOD FUEL USE IN THE FOREST PRODUCTS INDUSTRY

R. L. Jamison

Division of Energy Management
Weyerhauser Company
Tacoma, Washington

I.	INTRODUCTION	28
II.	ASSESSMENT OF WOOD FUEL QUALITY	28
III.	CURRENT USE AND SUPPLY TRENDS	33
	A. Manufacturing Residuals	33
	B. Forest Residuals	35
IV.	ENERGY RECOVERY TECHNOLOGIES USED IN THE FOREST PRODUCTS INDUSTRY	40
	A. Hog Fuel Combustion Technology	40
	B. Cogeneration	44
	C. Conversion Technologies	48
V.	INSTITUTIONAL ISSUES	48
VI.	CONCLUSIONS	50

I. INTRODUCTION

Until 1900 wood furnished a high percentage of U.S. energy supply, whereas today only 2+ percent of this nation's energy comes from wood. The trends in wood fuel utilization are now being reversed, led by significant activities in the forest products industry designed to make that industry more and more energy self-sufficient. Now that there is a greater demand for nonpetroleum energy, increased incentive exists to convert unproductive forest lands into forests of higher economic productivity. Millions of U.S. acres which are currently stocked with scrub and cull biomass have such little economic value that nothing can be done within the current price framework. Energy production may, given a proper economic climate, defray the costs of these stand conversions while contributing to the forest product industry's fuel production system.

This concept of garnering residuals from the forest is of particular importance to the forest industries, since they are the primary users of wood fuels. Further, with the supply of manufacturing residues being all but committed, any additional wood fuel increments of significant proportion must come from the forest itself.

II. AN ASSESSMENT OF WOOD FUEL QUALITY

Wood is not an ideal fuel. It is solid, of modest heat content, frequently wet, and bulky to transport. Wood fuels have certain advantages, however. Wood is a renewable resource. The sulfur and nitrogen contents in the various species of wood are lower than the sulfur and nitrogen contents of most coals and oils. Wood flyash may be an aesthetic issue, but not a health hazard; wood is a cleaner fuel than coal. This cleanliness is particularly significant in the forest products industry in meeting air pollution requirements. Many sawmills and other processing facilities are in or near Class 1 areas as defined by the U. S. Environmental Protection Agency (EPA).

Table I, taken from Junge (1975) presents proximate and ultimate analyses of hogged Douglas-fir and Western Hemlock bark on a moisture free basis. It shows that two of the principal sources of wood fuels are, indeed, environmentally superior to the coals available today.

TABLE I. Analyses of Selected Fuels
(Moisture Free Basis)

	Douglas-fir bark	Western Hemlock bark
Proximate analysis		
Volatiles	70.6 - 73.0	74.3
Charcoal	25.8 - 27.2	24.7
Ash	1.2 - 2.2	1.7
Ultimate analysis		
H	6.2	5.8
C	53.0	51.2
O	39.3	39.2
N	0.0	0.1
Ash	1.5	3.7
Higher heating value	10,100	9,800

Wood also has certain disadvantages. As has been previously mentioned, and demonstrated in Table I, wood fuels are modest in heat content and highly oxygenated. Being a solid fuel harvested above ground, wood can be messy to handle, shedding debris and dirt from bark crevices during the fuels handling processes. Fine wood is dusty, creating an explosion hazard unless properly handled. Wood ash requires careful disposal. The consequence of these problems is that wood systems cost more and have lower operating reliability than oil and gas systems, as later sections of this paper will discuss.

In addition to the dirt and dust problems, wood fuels available to the forest industry from manufacturing operations suffer from the following problems: (1) extreme variability in moisture content, hence variability in net heat available; and (2) significant variability in particle size, hence surface to mass ratio and combustion rate. Figure 1 is a pictorial representation of hog fuel while Figure 2 shows its sources in a sawmill. It is clear, particularly from Figure 2, why variability exists. What Figure 2 does not

FIGURE 1. A typical hog fuel pile. The photograph above shows typical wood fuel composed of bark, sawdust, shavings, and various other materials. It is variable in particle size, moisture content, and percentage of inerts. Its solid nature, plus its variability, causes large investments in materials handling equipment as this figure also shows. (Source: The Weyerhauser Co.)

show is that variations occur from hour to hour and day to day, both in moisture content and particle size, as the level of activity in any given machining center is increased or decreased. Thus hog fuels may vary in fine materials content from 20% to 60%, for example. The sources of hog fuel are presented in Table II, along with typical moisture contents and particle sizes.

There has been a substantial trend in the industry over the past several years to channel clean dry manufacturing residuals into raw material uses. Coarse wood residues are chipped and sent to pulp mills. Now some sawdust is also being pulped, rather than burned. The implication of this trend is that the wetter, dirtier materials are relegated to

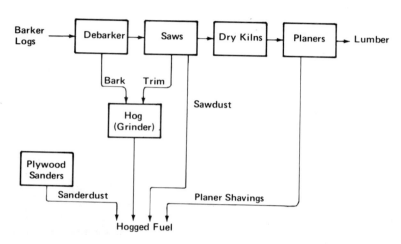

FIGURE 2. Sources of lumber mill manufacturing residuals. This figure shows in simplified form, the in-plant sources of hog fuel. From it one can visualize causes of variability. Barker logs coming from a dry yard will have lost some of their inherent moisture. Logs from a log pond will have gained additional water content. Debarkers may be either mechanical or hydraulic, again contributing to variable fuel quality. The primary saw, the headrig, may be of band saw design producing coarse sawdust. The headrig may be a chipping type of operation, producing some fines but no sawdust. Resaws, ponys, edgers and trimmers all provide room for fuel quality variability. The extent to which an operation is an integrated mill also affects fuel quality, as reflected in the contribution of sanderdust.

Fuel particle size, then, depends upon where in the process train the fuel comes from, the specifics of that process train, and whether or not it is hogged or ground.

TABLE II. Types and Values of Mill Processing Residuals

	% MC^a	Size^b
Bark	40-65	Coarse
Sawdust	40-55	Fine
Planer shavings	15-20	Coarse
Sander dust	7-15	Very fine
Chip fines	40-60	Fine
Trim, clippings, cores	15-60	Coarse
Clarifier sludge	60-90	Fine
Fly carbon	0	Fine
Log yard debris	40-70	Coarse and fine

^{a}Green wgt basis.
^{b}Coarse = $\sim$1/2"-4"; fine = $\sim$1/16" - 1/4"; very fine $\cong$ <1/32".

energy uses. The last three materials listed in Table II are not energy sources at all; rather, they are wastes to be disposed of by incineration. Such wastes, particularly clarifier sludge, are net energy consumers rather than energy producers and are burned when no better disposal solution exists. The net result, then, of this trend is the building of more costly systems in order to burn what wood fuel we have efficiently and cleanly.

The economic penalties to the forest products industry associated with fuel variability and degradation in quality go beyond the requirement for more expensive energy recovery systems. In order for these fuels to be burned cleanly, systems must be operated to handle the fuel component with the highest moisture content and the lowest combustion rate regardless of the penalties paid in reduced energy recovery efficiency and increased operating costs. Handling these fuels, for example, involves increasing the excess air admitted to the combustor in order to carry off the moisture as vapor; and such practices incur significant energy losses.

III. CURRENT USE AND SUPPLY TRENDS

The main source of wood fuel is spent pulping liquor, which is almost totally used to produce steam. Because of its high chemical content it must be recovered and recycled, and the Tomlinson Furnace has been the most efficient system for combined energy and chemicals recovery yet commercialized. This source of energy accounts for some 0.7 quadrillion Btu (quads) of annual energy supply to the pulp and paper industry. Manufacturing residuals from sawmills and plywood mills, as discussed above, are the next most readily available energy source, but that supply is nearly totally committed already, and is diminishing in quantity as forest industry processes are improved. The manufacturing residues excluding spent liquor now supply the forest products industry with 0.5 quads of fuel. Despite their declining availability and quality, their use is of considerable importance. Forest residuals—logging residues and noncommercial stands—are the next increments of wood fuel supply.

A. Manufacturing Residuals

Table III shows the supply pattern for manufacturing residuals in the Pacific Northwest states. Whereas the Pacific Northwest accounts for about 30% of total U.S. residuals generation, only 10% of this presently goes unused. Oregon generates the largest quantities of manufacturing residuals, and has the highest degree of residual utilization in the region. Since manufacturing residues are readily available and are disposal problems if not used, they are an integral part of raw material and energy supply now, as the data from Oregon shows. Estimates of national availability of manufacturing residuals show a 76% utilization rate. Table IV displays sources and volumes of wood residuals from manufacturing in Washington state forest industry operations. Of the 6.1 million oven dry tons (ODT) produced annually, 52% are used for furnish or materials production, 38% are used for fuel, and 10% are not used at this time.

In the forest products industry the sawmills are the largest source of hogged fuels and residuals, and the pulp mills are the largest users. Balancing these types of production operations in the northwest and the southeast helps account for the high rates of utilization. This high rate of utilization is reflected in the level of energy self-sufficiency which the forest products industry enjoys.

TABLE III. Mill Residuals in the PNW - 1976
(Millions of Dry Tons/Year)

	Residuals Production		
	Wood	Bark	Unused wood & bark
Pacific Northwest			
Oregon	12.2	3.0	0.5
Washington	4.8	1.3	0.5
Idaho / W. Montana	3.7	1.0	1.6
Alaska	0.2	.05	0.1
Total	20.9	5.35	2.7
Total U.S.	82		20.0

TABLE IV. Washington State Wood and Bark Residuals - 1976^a (Millions of Tons, Dry Wgt)

Source	Total produced	Used for furnish	Fuel use	Unused
Sawmills	4.6	2.6	1.7	.34
Plywood and veneer	1.1	0.6	0.5	.05
Shake and shingle	0.2	0.0	0.08	.16
Pulp and paperboard	0.03	0.0	0.03	0
Total	6.1	3.2	2.3	.55

^{a}Excluding black liquor solids from pulp mills.

TABLE V. The Forest Products Industry Use of Wood Energy (in Quads)

	1972	1977
Purchased fossil fuels and electricity	1.6	1.5
Wood wastes (and pulping liquor)	1.1	1.2
% from waste wood	41.0	45.0

As Table V shows this level has increased from 41% to 45% in the past five years, with the forest industries now supplying 1.2 quads of their 2.7 quad energy requirement. Because of this energy supply system, only 2.5 million ODT of manufacturing residuals go unused in the northwest, and only 20 million ODT of manufacturing residuals remain uncommitted in the nation as a whole.

B. Forest Residuals

The declining availability of hog fuel, given its importance in the forest industry, implies that the next increment of fuel which is forest residuals must be developed in the near future if the forest products industry is to continue its drive for increasing energy self-sufficiency. Forest residuals include both logging residues and nonmerchantable timber on lands which could and should be converted into more productive forestry.

The next increment is logging residuals of which there are roughly 11 million ODT/yr available in the Pacific Northwest and 100 million ODT/yr generated nationwide (Bethel et al., 1979). This latter figure corresponds to approximately 1.5 quads of energy, or the amount of energy currently being purchased by the forest industries. Logging residuals will not become significant in the energy picture, however, until more economic harvesting and transportation technologies are developed. Further, longer term, higher value raw material uses will compete for these wood residuals, reducing the supplies available for fuel purposes. An appropriate solution to this dilemma is to expand wood energy supply from the forests—going beyond the logging residues themselves.

This can be done by more intensive management of the forest resources we now have.

Yields from our national forests are considerably lower than from forestry lands. National forests typically produce 30 ft^3/acre/yr compared to 52 ft^3/acre/yr for industrial forestry lands. The forest industry, with 13% of the U.S. commercial forest land base accounts for 35% of the national production of timber on a growth basis. This difference can be corrected by planting and restocking millions of acres of public and private forest lands in highly productive species. Net annual growth from the 300 million acres of commercial forest land in farms and private ownerships of small to medium sizes, now producing 36 ft^3/acre/yr could improve under intensive management regimes. Growth per acre could be doubled on the less productive forest lands.

Neither the logging residues nor the materials from more intensive management practices will be inexpensive sources of fuels and energy. Regional and national studies indicate that, following the mill residues, logging residues will be the lowest cost fuels. Materials from stand conversions and stand improvements will then follow. There is, of course, considerable overlap in these fuel sources as costs are dependent upon material accessibility and amenability to mechanized harvesting systems.

Figure 3 is a marginal cost step function which shows that the unit cost of wood fuel goes up as the quantity delivered to a specific point is increased. These data were obtained from tests made in Weyerhauser's Oklahoma Region. The solid lines depict average costs of delivering successive increments of wood from specific stands within a 100 mile radius, although the average haul distance is only 30-50 miles. The stands surveyed vary widely from large diameter hardwoods to smaller diameter mixed hardwoods and softwoods of different ages, stocking densities, terrain, and distance. One increment is simply tops and limbs from harvested stems—the most expensive form of logging residues. The dotted line depicts expected costs as larger quantities are sought from greater distances and rougher terrain. The cheapest wood is that which is most accessible to roads, is located nearest the point of use, and is in big land areas on flat terrain suitable for mechanized harvesting and processing into fuel.

Each successive block in Figure 3 averages in a higher cost increment than the one before, thus forcing up the average cost of all wood fuels as the volume delivered increases. The dotted line of $27/ODT is a benchmark indicating the value of wood fuel equivalent to oil at $13/bbl. Increments in the lower left of Figure 3 are large diameter hardwoods in a relatively dense stand on flat terrain close

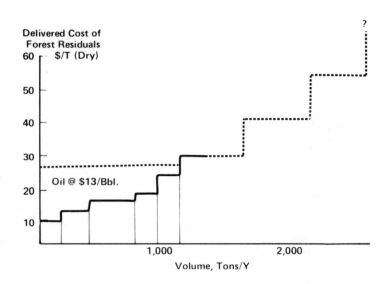

FIGURE 3. The price supply relationship for wood fuels. This figure is a step function type supply curve for wood fuels delivered to a specific Oklahoma mill owned and operated by Weyerhauser. The higher cost increments include small pieces, wood located relatively distant from the plant, and wood not amenable to mechanized harvesting and transportation. Supply curves do not vary in general shape, but vary in slope, as a function of geographical area based upon terrain and infrastructure.

to the pulp mill in Oklahoma used as the focal point of the study. The cost of bringing in these hardwoods is less than $20/ODT. Only a small quantity of material fit that set of criteria, however. At the other end of the solid line the increments are small pieces of logging residues costing as high as $67/ODT to harvest and deliver. This more expensive material raises the average wood fuel cost from less than $20/ODT to $30/ODT.

Oklahoma has no forest residuals flowing to the plant for energy purposes at the present time. In this area lower cost manufacturing residuals are still available in sufficient quantity to satisfy the plant's wood fuel system. In North Carolina, however, the situation is somewhat different. There the terrain is flatter and the stands are more amenable to mechanized harvesting with its attendant economics. Early in 1978 Weyerhauser began augmenting its hog fuel supply with forest residuals; currently 15% of the fuel stream in that situation comes from the forest.

The costs of obtaining wood fuels from various sources, at the national level, can be represented by ranges for manufacturing residuals, forest residuals, and fuels from energy farms or fuel plantations. These ranges are presented here in Figure 4. Further they are compared to existing fossil fuels such as imported oil and natural gas. Figure 4 clearly shows that manufacturing residues and some forest residuals are economic today. From a production cost if not an opportunity cost point of view, specifically grown fuels may become a consideration in the not-too-distant future.

In summary, then, considerable additional energy could be obtained from logging residues although the cost of doing so in a large way would be significantly higher than could be justified within the constraints of today's oil prices. Some of this energy is now being produced as it becomes economic. More will have to be produced if the forest products industry is to improve its energy self-sufficiency position. As Table V showed, the forest industry has made considerable gains in that regard. The forest products industry has the wood supply, has the experience in using wood energy, and has a need to replace the 1.5 quads of energy presently being purchased. It is incumbent upon this industry to continue the trend towards energy self-sufficiency to displace petroleum fuels and make them available to others less able to switch to wood.

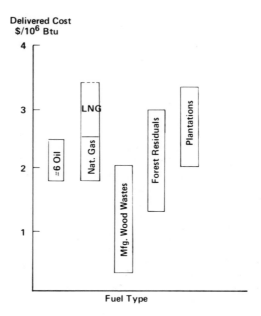

FIGURE 4. *The cost of obtaining various fuels. Figure 4 shows that, given the current price ranges for oil and natural gas, virtually all manufacturing residues are economically available as fuel. Some fraction of forest residuals are also available, although the size of that fraction is modest. Today 15% of the fuel requirement for one North Carolina mill, for example, comes from this source. Finally some estimates show that plantation grown fuels can be produced at the $2/10^6 Btu level. Such costs do not include opportunity costs for using the land in such a manner, however. As fossil fuel prices continue to rise more use of wood fuel is expected, with emphasis on the forest residuals.*

IV. ENERGY RECOVERY TECHNOLOGIES USED IN THE
    FOREST PRODUCTS INDUSTRY

There are several technologies important in the utilization of wood for energy. Conventional technology is typified by the hog fuel boiler, which comes in many designs and sizes. Over the past several years there have been significant gains in hog fuel boiler technology, mostly of an evolutionary nature. These gains will be discussed below. Hog fuel boilers are employed typically to raise process steam. Cogeneration has been employed frequently as an extension of hog fuel combustion technology, allowing the forest industries to raise a portion of their electricity requirement along with steam. This will also be considered in the ensuing paragraphs.

A. *Hog Fuel Combustion Technology*

Hog fuel boilers have increased in size from the 15,000 lb/hr dutch oven units of the 1920s to 500,000 lb/hr traveling grate giants of the 1970s. Modern wood fueled boilers use mechanical draft fans to replace the tall stack natural draft systems originally employed. Spreader stokers distribute the fuel evenly over large grate areas to ensure complete and efficient combustion. This contrasts to the pile burning systems associated with dutch oven technology. With the traveling grates, ash can be removed continuously, eliminating the hand rake systems. Furnace construction has changed from the squat, refractory lined chambers of the dutch oven to the tall waterwall designs which increase energy recovery efficiency and prolong refractory life. Finally, boiler instrumentation and combustion controls are vastly improved. Air flows are metered, fuel-to-air ratios are controlled, and combustion efficiency is monitored by oxygen and carbon monoxide concentrations in the flue gas. Stack emissions are monitored by in-stack opacity measurement.

These developments have improved both energy recovery and environmental acceptibility of wood fueled systems in the forest products industry. The disadvantage is that the increases in complexity result in lower system reliability unless very high quality and/or redundancy are provided. A corollary disadvantage is the high cost of such systems. Figure 5 is a photograph of a modern hog fuel boiler designed for maximum efficiency when operating on variable quality and high moisture wood fuels. It is located at the

FIGURE 5. The Weyerhauser energy recovery system of Longview, Washington. The system shown above is the 500,000 lb/hr Longview boiler, the largest wood-fired facility built to date. It not only produces process steam but also is a cogeneration facility.

Weyerhauser operations in Longview, Washington, and has a capacity of 500,000 lb/hr of high pressure steam. It is the largest such system built to date.

The deterioration of hog fuel quality discussed in previous paragraphs makes it more difficult to obtain efficient combustion, high reliability, and low stack emissions simultaneously. Three approaches to this problem are available. First, the fuel can be processed before combustion to remove dirt and moisture, and produce a clean burning, efficient fuel. A second alternative is the development of a combustion system with the capability of burning even the lowest quality fuel in a reasonable manner. Finally, systems can be added on to conventional boilers to increase energy extraction efficiency and control emissions. All three of these basic approaches are being pursued today. Each is appropriate in individual situations. While all three approaches are being pursued, it is clearly better to solve a problem than to treat the symptoms or results. Thus the first two approaches of fuel quality upgrading and new system design are preferable to adding more complexities to existing equipment.

The processes applicable to upgrading the hog fuel are drying, screening, grinding, washing, and pelletizing or densification. The objective of these processes is the removal of noncombustible dirt and moisture, reduction of oversize and slow burning material, and agglomeration of undersize or fine material. The end result sought is a clean, dry, uniformly sized fuel. The uniformity of size is important for ease in transportation, storage, and fuel feeding. Reduction of dirt and moisture content reduces transportation costs, promotes efficient combustion, and minimizes air emissions. These systems address the problems of requiring large quantities of excess air as previously discussed.

A typical hog fuel system now includes screening and grinding or hogging of oversize material. If the fuel is unusually dirty it may be washed to remove dirt, with losses in thermal efficiency being incurred. Washing is not a common practice.

Drying may be employed on washed or nonwashed fuel, and may be accomplished either with flue gas waste heat or by burning some of the fuel being dried. The gain in system efficiency is relatively small, however. The principal advantages of drying are increasing the capacity of a given boiler and reducing the total capital investment of the system. A reduction in moisture content of the fuel from 50% to 30% (green wgt. basis) has been shown to increase the capacity of the boiler by 25% in terms of steam output. This capacity increase is particularly significant if a company is considering adding a second boiler to an existing energy recovery plant. High moisture content finely divided wood fuel is most amenable to drying. If flue gas waste heat is to be the heat source, economics will be favored by high temperatures (e.g., 700-1000°F).

One additional stage of fuel preparation often proposed is pelletizing, or forming fuel into uniformly sized and densified particles. This process can reduce transportation and handling costs, but for larger scale operations the gains are usually not economically beneficial.

The second approach involves changing the combustion process in order to get complete fuel burnout with a minimum efficiency loss and a minimum of stack emissions. This can be achieved with either improved combustor design or fuel modifications, or combinations of the two.

Examples of improved combustor design are the fluidized bed burners and the two cell pyrolytic burners. The fluidized bed combustors obtain excellent fuel burnout at relatively high efficiencies with low quality, nonuniform fuels. The pyrolytic burner (e.g., the Lamb Cargate wet cell burner) has very low stack emissions with relatively high thermal efficiencies and fuel burnout rates. A third type of burner is the cyclonic vortex system such as the Energex.™ Cyclone burners require high quality fuels (e.g., fuels with less than 15% moisture and low ash content) which are finely divided. These systems, given good fuels, have been very successful in direct fired kiln applications and plywood veneer drying systems within the wood products industry. They are specialized systems for given applications, however.

Until the above approaches become operationally feasible on a more universal basis, and become economically attractive for more than specialized situations, the only remaining alternative is environmental controls applied to the stack. Such controls fall into two groups, wet and dry. The wet types produce dense white plumes of water vapor and a small volume of contaminated water which must be treated before release. Wet scrubbers are typical of this group. Wet scrubbers may be low, medium, or high energy consumers with proportionate fan requirements. Dry control systems are baghouses, electrostatic precipitators, moving gravel bed dry scrubbers, and mechanical cyclones.

As the above discussion shows, wood-fired systems are complex entities. Typically the capital cost of such systems is between 3 and 4 times the capital cost of a comparable petroleum-fired installation. Today as much as 25% of that investment is in fuel handling, getting the wood into the boiler. Figure 1, the hog fuel pile, shows some of the materials handling equipment required. Another 20% is in the air pollution control system. Thus wood fired systems depend on low cost fuel supplies—particularly in light of the reduced efficiency of burning wood when compared to oil or natural gas (typically 68% thermal efficiency for wood and 82% thermal efficiency for oil and gas). Given these cost and efficiency differentials, the significance of the supply step functions presented earlier can easily be seen. Further, when one considers that on one boiler of 100,000 lb/hr capacity a reduction of particulate emissions from 0.5 lb/10^6 Btu to 0.25 lb/10^6 Btu would have required an additional investment of $2 million (Methven, Jamison, and Shade, 1978), the need for system improvements becomes quite clear.

The forest products industry does have useful technologies for recovering energy as process heat or steam. These technologies are cost effective in the context of the present fuel supplies, and emerging technologies promise greater economic benefits. Improvements are needed, however, if the forest products industry is to become less dependent on purchased fuel by going into the forest for residues.

B. *Cogeneration*

Cogeneration is another wood energy technology widely employed in the forest products industry. Cogeneration is the concurrent generation of electricity and the use of exhaust heat (usually in the form of process steam) for manufacturing operations. This is accomplished by burning fuel to make high pressure (e.g., 600-1200 psig) steam, passing this steam through a back pressure or extraction turbine driving a generator, and then using the steam exhausted from the turbine at lower pressures (e.g., 50-250 psig) for process heat. This technology gets full use of the energy contained in the fuel. In contrast, a utility turbine passes the exhaust steam to a condenser where over half of the fuel energy is wasted. With wood fuel at 55% moisture, power can be generated at an efficiency of about 60% compared to modern utility turbines with efficiencies of 33% when using nuclear fuels and 39% when using fossil fuels. Expressed another way, the heat rate for cogeneration with hog fuel is 6000 Btu/kwh compared to 9500 Btu/kwh for a fossil burning utility system and 15,000-20,000 Btu/kwh for a condensing turbine fueled with wood.

The cogeneration concept also applies when the turbine directly drives process equipment rather than generates electricity which in turn powers machinery motors. A third example of cogeneration is the use of exhaust heat from a gas turbine rather than the exhaust steam from a steam turbine.

Cogeneration is commonly deployed throughout the forest products industry. For example, the Weyerhauser Company has eleven cogeneration facilities (22 turbine generators) in operation in the United States producing both electricity and process steam. In addition we have 25 plants that use steam turbines as mechanical drives rather than producing more electricity for a higher plant power load.

Most of these cogeneration facilities are back pressure turbines. They are supplied with high pressure steam (and perhaps with extraction ports), and then exhaust steam at a lower pressure sufficient to satisfy process requirements. As examples of the better cases, several Weyerhauser plants are generating about 60% of their electricity requirement and one integrated plant is generating 87% of its electricity requirements. Generation by back pressure systems is popular, but opportunities are limited by the steam demand in a plant. Relatively large steam requirements are necessary in order to support such turbines (Methven, 1978).

The forest products industry record with respect to cogeneration is better than other types of manufacturing in the U.S. Table VI provides salient data for the Pacific Northwest; Tables VII and VIII provide data for the nation as a whole. In Table VII the pulp and paper industry represents the forest industries. Significant to note from Table VIII is the fact that on an input basis, organic wastes are the most prominent fuel identified by Resource Planning Associates (1977). They expect this to continue through 1985. However, there is considerable room for improvement.

To a large extent the employment of cogeneration systems depends upon utility pricing and attitude at the local or service area level. Rapidly increasing prices of industrial electricity have meant that, where once 25 megawatt (MW) systems were the minimum economic size, today systems in the 3 to 5 MW range may be economic on a site specific basis. The importance of utility attitude can be seen from the cooperation between the Weyerhauser Co. and the Eugene Water and Electric Board (EWEB) in Eugene, Oregon. The example involves the construction of the "Utility Industrial Energy Center" in Springfield, Oregon. There the Weyerhauser boilers associated with a pulp mill are producing 1.2 million lb/hr of 850 psig steam. EWEB leases a small site from Weyerhauser at the pulp mill where EWEB has installed a 50 MW cogeneration turbine. The Weyerhauser steam is passed through the EWEB turbine before going to pulp mill processes, and EWEB pays for the high grade energy in that steam. EWEB is able to generate electricity at a cost of about 2¢/kwh (Hunt, 1978). This results from not having to own and operate its own boilers, and not wasting 67% of the energy in turbine exhaust.

In summary, because the forest products industry requires large quantities of process steam there is an opportunity to cogenerate and, in fact, the industry does produce about 50% of the electricity it uses. In theory, many current pulp and paper mills could be electrically self-sufficient, although this is not commonly done, and improved steam utilization makes this prospect increasingly unlikely. Further, a standby tie with a utility is desirable to provide backup power or system reliability in case the mill power generation system fails. The cost of the standby tie-line makes it economic to purchase a small amount of power from the utility. Thus electrical self-sufficiency rates of 80% to 90% are more reasonable to forecast. There is also a minimum size plant for economic power generation. Unless steam usage is more than 70,000-120,000 lb/hr, equivalent to 3-5 MW of back pressure, it is not likely to be economically viable to deploy cogeneration systems.

TABLE VI. Cogeneration Capacity Employed in the Pacific Northwest

Industry	Cogeneration capacity employed (MWe)[a]
Forest based industries	
Pulp and paper	120[b]
Plywood	31
Sawmills	17
Total	168
Other industries	
Food processing	19
Total	19
Total	187

[a] Megawatts electric.

[b] This does not include 50 MW of capacity owned by the Eugene Water and Electric Board (EWEB) described in the text. When added in the total is 237 MWe.

Source: Rocket Research (1979).

TABLE VII. Cogeneration Activity in Six Major Industries in the U.S., 1976.

Industry	Energy in process steam used in cogeneration		Electricity generated by cogeneration	
	10^{12} Btu	%	10^9 kwh	%
Pulp and paper	236	40.4	10.14	32.3
Chemicals	190	32.5	11.44	36.5
Steel	101	17.3	7.00	22.3
Food	27	4.6	1.16	3.7
Petroleum refining	15	2.6	0.94	3.0
Textiles	15	2.6	0.69	2.2
Total	584	100.0	31.37	100.0

Source: Resource Planning Associates, 1977.

TABLE VIII. Steam and Electric Power Produced by Cogeneration as a Function of Fuel, 1976

Fuel	Gross quantity used (10^{12} Btu)[a]	Steam generated (10^{12} Btu)	Electricity generated (10^9 kwh)
Organic wastes[b]	360	182	9.90
Residual oil	282	204	8.81
Coal	136	97	4.16
Other	150	101	8.5
Total	928	584	31.37

[a] Calculated based on steam and electricity produced.
[b] Primarily hog fuel and spent liquor.
Source: Resource Planning Associates (1977).

C. Conversion Technologies

As an alternative to combustion, wood fuels can be converted into fuel forms suitable for existing oil and gas combustor designs. The two technologies being developed are wood pyrolysis and wood gasification. These two processes are similar, differing in the method and extent of heat application. Pyrolysis (see chapter by J. A. Knight) in the pure sense is indirect heating to thermally decompose the wood. Pyrolysis may also be accomplished by partial oxidation at very low levels. Depending upon temperature, various combinations of three products are produced: noncondensible fuel gas, condensed organic oils, and solid char. Gasification involves addition of air for partial oxidation of the char into carbon oxides, and exothermic process which supports the remaining gasification reactions. While both processes convert solid wood into one or more clean fuel fractions which are easier to use than solid wood, they do not produce fuels which are directly substitutable for the natural gas and fuel oil already in use. However, low cost technology is available for efficient, clean combustion of these fuels either separately or in combination. Overall thermal efficiency of gasification and subsequent combustion of the fuel gas produced can be higher than for direct wood combustion. The main advantage of these energy conversion technologies, however, is to enable continued use of existing combustion equipment while using wood fuels. Retrofit to existing investments is an essential development for rapid increase in wood fuel use.

V. INSTITUTIONAL ISSUES

Some of the barriers to greater use of wood for energy include insufficient rate of return on wood fueled systems, low fossil fuel prices making wood uneconomic to burn, high costs of harvesting and transporting forest residuals, and stringent air emission standards which increase the cost of wood fired systems by as much as 25%. As has been mentioned previously, wood burning systems have capital costs 3 to 4 times greater than oil and natural gas systems of the same capacity. Additionally pollution controls on wood burning systems are more costly than those for oil and gas systems; and wood fueled systems, being more complex, have lower reliability. Consequently wood fueled systems are built with more redundancy, extra quality, and more capacity for insurance purposes. It is substantially more difficult, therefore, to justify solid wood fueled systems in economic terms.

There are three approaches to these barriers which provide solutions to the problem of economic viability: (1) higher energy prices increasing the savings which accrue to a company switching to wood fuel and generating a higher rate of return on the capital investment (see chapter by D. A. Tillman); (2) lower delivered cost of forest residuals used as fuel; and (3) energy recovery technologies with lower capital costs for wood handling, burning, and environmental controls.

Historically, efforts designed to pursue the third approach have been fruitless as energy systems have continued to become more complex and more costly. We can hope for new technology that will cost less, but in my experience that may not occur. Thus the first and second approaches must be addressed most vigorously while the third is not ignored.

Based on the above, incentives are needed that would improve the short-term economics for conversion/replacement of petroleum burning equipment to or with wood-fired systems. Table IX lists some suggestions made recently to the Department of Energy for accelerating the use of wood energy. The first group is aimed at providing a two- to three-year payback on wood burning equipment—the payback period required in order to get these facilities built quickly. In one form the incentive could be an additional 20% or more of investment tax credit on new energy-related wood-using equipment. Another incentive would be waiving the 50% ceiling on investment tax credit utilization. A third, but less effective alternative, is accelerated depreciation. Another assist would be revised air emission standards for wood burning equipment to reduce the investment cost of controlling emissions to unnecessarily low levels. Current standards were derived from coal burning practice, which has a set of problems which are quite different from those associated with wood. Wood has almost no sulfur, and recent tests indicate that flue gases are neither mutogenic nor carcinogenic hazards to health. Therefore, it makes sense to develop specific wood burning standards ensuring that emissions are not a nuisance.

The second group of incentives identified in Table IX is aimed at increasing wood fuel supply in the longer term. Foremost is the need to improve the climate for private investment in the forest resource itself. Intensive forest management practices must be made economic and attractive for all owners—public and private. The Forest Service must speed up its harvest rotations, shortening the cycle by

TABLE IX. Incentives Appropriate for Increasing the Use of Wood Fuels

Incentives for wood utilization facilities
- Additional 20-30% investment tax credit
- Waive 50% limit on tax credit utilization
- Accelerated depreciation
- Appropriate environmental standards for wood fueled facilities

Incentives for increased supply
- Reducing rotations and restocking in national forests
- More intensive forest management
- Continuing R & D for improved genetics and silviculture
- Improving climate for private investment in forest resource

---

10 to 30 years, and it must increase its forest stocking densities. Government research and development support for improved tree genetics and silvicultural methods is also needed.

VI. CONCLUSIONS

The above discussion presents a forest industry perspective on use of wood fuels. Clearly the fuel is of modest heat content, has the advantages of being renewable and low in pollution potential, and has the disadvantages associated with a high moisture content. Supplies of wood fuels are available now in limited quantities, and the forest products industry is able to obtain 45% of its energy, or 1.2 quads, from wood fuels. If that percentage continues to increase in ensuing years, however, fuel must be brought in from the forest at costs higher than those associated with current wood fuels.

Technologies for combustion and cogeneration now exist to convert wood fuels into useful energy. These technologies are more expensive and less efficient than comparable oil and gas fired equipment and depend on low cost fuel to be economic. If wood fuel utilization is to be expanded the institutional barriers associated with the high cost of owning and operating wood fueled systems must be replaced with incentives for wood fuel utilization. These incentives are particularly important if the forest products industry is to decrease its reliance on purchased fossil fuels, making those fuels available to sectors of the economy which have no opportunity to convert to wood or other biomass sources.

REFERENCES

Bethel, J. S. et al. (1979). "Energy from Wood." A Report to the United States Congress, Office of Technology Assessment. College of Forest Resources, University of Washington, Seattle, Washington.

Hunt, Herbert H. (1978). Utility industrial energy center, in "Increased Energy from Biomass: 1985 Possibilities and Problems, Proc." Pacific Northwest Bioconversion Workshop, Oct. 24-26, Portland, Oregon.

Jamison, Robert (1978). Manufacturing residue for energy, in "Increased Energy from Biomass: 1985 Possibilities and Problems, Proc." Pacific Northwest Bioconversion Workshop, Oct. 24-26, Portland, Oregon.

Junge, David (1975). "Wood and Bark Fired Boilers." Oregon State University Res. Bulletin No. 17, Corvallis, Oregon.

Methven, Norbert E. (1978). Existing cogeneration facilities, in "Increased Energy from Biomass: 1985 Possibilities and Problems, Proc." Pacific Northwest Bioconversion Workshop, Oct. 24-26, Portland, Oregon.

Methven, N. E., Jamison, R. L., and Shade, R. A. (1978). "Energy from Forest Biomass." A Report of Task Force #5 of the Industrial Energy Group, National Association of Manufacturers, Washington, D.C.

Resource Planning Associates (1977). Potential for cogeneration development, *in* "Cogeneration of Steam and Electric Power (Robert Noyes, ed.). Noyes Data Corp., Park Ridge, N.J., 1978.

Rocket Research Company (1979). "Cogeneration Potential in the Bonneville Power Administration Service Area, Phase 1." Bonneville Power Administration, Dept. of Energy, Portland, Oregon.

THE ECONOMIC VALUES OF WOOD RESIDUES AS FUEL

David A. Tillman

College of Forest Resources
University of Washington
Seattle, Washington

I.	INTRODUCTION	54
II.	BACKGROUND	55
III.	PARAMETERS AND DEFINITIONS	56
	A. Assumptions	56
	B. Limitations	58
IV.	THE VALUE OF USING WOOD AS A FUEL	59
	A. The Energy Value of Wood	59
	B. The Cost of Using Wood Fuel to Generate Steam by Direct Combustion	62
	C. The Opportunity Cost Curve for Wood Fuels Based on Combustion and Steam Generation	64
	D. The Technological Sensitivity of the Wood Fuel Opportunity Cost Curve	67
V.	THE VALUE OF WOOD RESIDUES IN THE MANUFACTURE OF MATERIALS	74
VI.	OPPORTUNITY COST COMPARISONS	76
	A. General Economic Value Curves	77
	B. The Influence of Rising Energy Prices on Wood Fuel Values	79
	C. The Influence of Transportation Costs on Residue Values	80
VII.	CONCLUSIONS	81

I. INTRODUCTION

Because of the growth in consumption of wood residues to fuel industry's boilers and kilns, potential competition for those residues exists between materials and energy applications. Chips can be used in pulp mills or furnaces; dry veneer trim and planer shavings can be sold to particleboard plants or burned; sawdust can be used by pulp mills, particleboard plants, or in boilers and kilns; and bark has minor non-fuel applications. Increasing attention is now being paid to the use of these mill residues as a source of energy, resulting in the raising of numerous economic questions. Among the questions raised are the following: (1) given the variability in fuel quality of wood residues particularly as influenced by moisture content, in what conditions are they more valuable as fuel and when are they more valuable as furnish; (2) how do changing energy prices affect the tradeoff between the use of residues as fuel and as furnish to pulp mills and particleboard plants; and (3) how do transportation costs (of residues used for furnish) influence the relative values of residues as fuels or sources of raw materials. To offer some answers to these questions, a series of opportunity curves have been constructed based on Pacific Northwest conditions.

In order to consider these issues, this paper will provide a brief background discussion and then perform the following:

(1) Consider the energy value of residues, particularly as it is affected by moisture content;

(2) Develop an opportunity cost/economic value curve for wood residues as a function of moisture content based upon direct combustion to produce process steam;

(3) Consider the influence of cogeneration and electricity generation by utilities on the economic value of residues;

(4) Describe the economic value of residues in traditional raw materials markets; and

(5) Compare the relative value of residues in materials and energy applications as a function of moisture content.

II. BACKGROUND

Industrial consumption of wood residues as a fuel source is growing rapidly through the U.S., establishing a competition for those materials between energy and materials applications. Estimating when wood is more valuable for its fuel or fiber content, therefore, is essential to forest and finished wood products companies and economists alike. Such an estimation is made somewhat complex by the variable nature of wood residues as fuels, and by the fragmented commercial markets for residues sold to supply heat or steam. Such an estimation, however, leads to conclusions that, depending on residue quality, oil price, and distance between residue producers and users; residues may be more valuable for energy than for materials purposes at current market prices.

Presently wood fuels are supplying 1.2×10^{15} Btu (quads) to industry (Tillman, 1977). As Table I shows, the vast majority of that energy is supplied to the pulp and paper industry, with lesser amounts being used in the lumber, plywood, furniture, and related manufacturing sectors. Table I presents such estimates. These estimates can be corroborated by estimates based on the national boiler inventory (Hall, 1976).

This residue energy supply represents a very large share of the total fuels budget for forest products industries; $\sim$45% for pulp and paper (Duke, 1978), $\sim$33% for sawmills, and $\sim$50% for veneer and plywood plants (Forest Service, 1976). For individual companies and plants wood residues play an even larger role. They supply over 60% of the energy

TABLE I. Wood Utilization as Fuel for 1976

User group	Wood and wood residue utilization (in 10^{15} Btu)
Pulp and paper	0.982
Sawmills, plywood mills, and veneer mills	0.070
Metallurgical industries	0.012
Other industries	0.105
Total	1.169

consumed by the Weyerhauser Co. in all of its plants (Jamison, 1977); and 98% of the energy for the Crown-Zellerbach, Omak, Washington operations (Furman, 1978). Boiler sales data also show growth in wood residue use for fuel purposes. During the 1960s, wood fueled boiler capacity as a percentage of total industrial boiler capacity sold averaged 2.2%, while it averaged nearly 9% of all industrial boiler sales in the 1970s (Tillman, 1978). Since 1970 wood fueled industrial boiler sales, including spent liquor recovery boilers, have about kept pace with coal fired industrial (non-utility) boilers sold on a capacity basis (Tillman, 1977). This growth is continuing in response to fossil energy availability, price pressures, and environmental concerns. Consumption is expected to be sufficiently high that Weyerhauser Co. officials Jamison (1977) and Christensen (1975), forecast turning to logging residues as a source of fuel and fiber in the foreseeable future.

Because of this growth in wood energy consumed, there have been several recent publications estimating the economic value of wood fuels. Such systems typically approach wood fuel value on a Btu for Btu replacement of coal and/or oil without considering the capital cost penalties associated with wood fuel systems. Frequently efficiency differentials are ignored also. A second approach has been used as well—equating value with the cost of production. Because of these systems, an opportunity cost system has been developed here.

III. PARAMETERS AND DEFINITIONS

This assessment of opportunity cost curves makes certain limiting assumptions; and employs certain definitions, as identified below.

A. Assumptions

The residues where competition is potentially most keen are planer shavings and sawdust. Pulp chips are useful for illustrative purposes. Since bark is not used in large quantities for nonfuel purposes, and is viewed as a contaminant by such operations as pulp mills, a fuel value but not a material value is presented. Energy values for residues will be presented primarily on an as-fired or total weight basis, and economic values will be expressed as both $/Green Ton and $/Bone Dry Ton.

For the imputed $/ton value of residues as an energy fuel, the common denominators used are lb. of steam generated, 10^3 Btu of direct heat raised, megawatts of electricity generated, or some combination thereof. While energy system size is recognized as a factor in determining value, economies of scale curves for all systems are essentially comparable for oil and wood fueled systems up to ~400,000 lb. steam/hr (Axtman, 1978). Initially 120,000 lb/hr systems are used since 33% of all wood fired systems are in the 50,000-150,000 lb/hr capacity range (Hall, 1976). Further, low pressure (150-300 psig) systems are employed initially since they are the most common in nonpower generation applications. The alternative fuel system employed for analysis is an industrial boiler burning heavy (e.g., #6) oil.

The systems employed for economic valuation are those which are presently commercially available: (1) combustion to raise process steam, with emphasis on spreader stoker systems; (2) cogeneration of steam and electricity in a system employing a back pressure turbine; and (3) electricity generation in a full condensing cycle, as operated by a utility. These systems present a variety of opportunities for using wood. They require somewhat different methodologies for opportunity cost of wood valuation.

While it might be convenient to use pyrolysis—pegging produced fuel values to comparable decontrolled natural gas and oil prices—there are three reasons for not doing so. The vast majority of wood residues are burned directly, with only a minor fraction being converted to charcoal or wood gas. Further, pyrolysis systems applied particularly to large particles (e.g., 1/2" size) introduce unnecessary complexities into the equation for economic purposes. Schlesinger (1972) has demonstrated that moisture influences not only the efficiency of conversion by pyrolysis, but also the quality of the gaseous fuel as a result of the water-gas shift $[CO + H_2O \rightleftarrows CO_2 + H_2]$ and steam-carbon $[C + H_2O \rightleftarrows CO + H_2]$ reactions. Further complexities come from variations in operating conditions (e.g., temperature, pressure) and system design (e.g., oxygen source, use or nonuse of catalysts) which make such comparisons difficult at best. Finally, such systems are not commercially available and in use at the present time.

The determination of economic value for residues assumes that a new fuel utilization system will have to be purchased either to burn oil or wood. From an investment point of view, one can also consider the options of installing a wood fired boiler vs. keeping an existing oil fired system—or vise versa. Such considerations are real in corporate capital budgeting, but outside the scope of this analysis.

TABLE II. The Cost of Capital for Wood Fueled Installations (Assume 25 year Life)

Item	Ave. % of initial plant investment	
	Industrial	Utility
Depreciation	4.00	4.00
Debt (12%/yr)	0.00	4.50
Equity (30%/yr)	15.00	3.75
Income tax (48%)	14.40	3.60
Total	33.40	15.85

The capital cost assumption is contained in Table II. It is the 100% equity assumption on an industrial plant with a 25 year life, and provides for a 3-year payback period. Such a short payoff is consistent with industrial practice. For utilities the capital cost assumption is quite different as is also shown in Table II. The 100% equity case assumes that industry demands the same performance from invested long term debt that it does from retained earnings. Using standard annuity tables provides a factor of .3501 for competitive industry and .1835 for utilities. The utility assumption is 25% equity and 75% debt financing. The utility case is consistent with existing practice. The market values for residues employed in nonfuel applications are delivered value for Douglas-fir residues as supplied by the Weyerhauser Co. (Jamison, 1978). Transportation rates employed are from the same source.

B. *Limitations*

The initial estimates presented below are generalizations. It is recognized that no generalization necessarily fits an individual case; values will deviate from the curves presented as companies vary in timber holdings, product mix, residue generation and tax incidence. For example, Weyerhauser is 100% self-sufficient in timber while Boise Cascade is only 25% self-sufficient (Quirk, 1977). Bohemia, Inc. is not involved in pulp and paper, the large residue consuming industry; other large firms are. Values will also vary by distance from organized residue markets as a function of transportation. These generalized curves can be modified, as needed to fit individual plant situations.

IV. THE VALUE OF USING WOOD AS A FUEL

Employing Douglas-fir as the species under consideration, one can examine the value of wood fuel, in terms of oil savings, in a threepart analysis: (1) describing the energy value of Douglas-fir, (2) defining the dollar costs of burning appropriate residues, and (3) estimating dollar savings obtained from not burning oil. Since fuel quality is defined primarily by chemical composition rather than physical characteristics (Shafizadeh, 1976), values are equally appropriate for chips, shavings, and sawdust. Bark is treated separately.

A. The Energy Value of Wood

The higher heating value of Douglas-fir wood, 9540 Btu/lb has been presented numerous times, including in papers by Arola (1976), Corder (1973), and others. Similarly moisture content values and influences have been discussed by Corder (1973), Riley (1976), Ince (1977), and numerous others. The list of references on this subject is legion. For purposes of subsequent analysis, gross heat available and net energy delivered as steam as a function of moisture content, are summarized in Fig. 1. The modified Ince formulae, used to develop Fig. 1, are as follows:

Heat available:

$$HA = (1 - MC) \times HHV \qquad (1)$$

Heat loss caused by moisture:

$$HL_M = MC \times [938 + (212 - T_1) + (0.253 \times T_2)] \qquad (2)$$

Heat loss caused by hydrogen:

$$HL_H = H \times (1 - MC) \times [938 + (212 - T_1) + (0.253 \times T_2)] \qquad (3)$$

Heat loss caused by dry gas and excess air:

$$HL_{dg} = (T_2 - T_1) \times [(0.796 \times \%EA) + 0.843] \qquad (4)$$

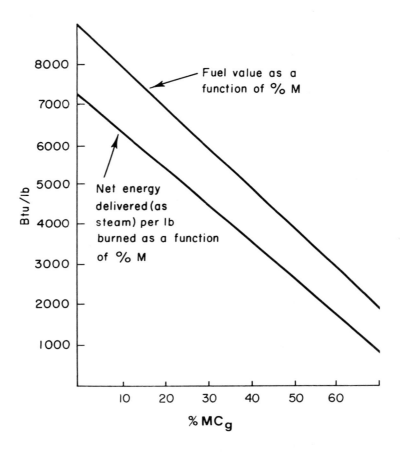

FIGURE 1. *The energy value of Douglas-fir residues as a function of moisture content.*

Combustion efficiency:

$$CE = \frac{HA - (HL_{M, H, dg}) + .04\, HA}{HA} \quad (5)$$

where:

- HA = heat available (Btu/lb)
- HHV = higher heating value (Btu/lb)
- MC = moisture content (total weight basis)
- HL_m = heat loss due to moisture (Btu/lb)
- T_1 = temperature of wood fuel before combustion (°F)
- T_2 = temperature of stack gas (°F)
- HL_H = heat loss due to hydrogen content of wood
- H = hydrogen content of wood (wt %)
- HL_{dg} = heat loss due to dry gas and excess air
- EA = excess air (%)
- CE = combustion efficiency

The values for Douglas-fir bark vary slightly from those of wood. The higher heating value, dry, is 10,300 Btu/lb (Shafizadeh, 1976). Combustion of bark is somewhat less efficient due to the sand and grit normally entrained in the bark fuel. The bark function of net heat delivered is, therefore, virtually identical to the wood function in the normal or average moisture range.

For comparative purposes, the higher heating values for selected other species are as follows: western hemlock, ∿8750 Btu/lb.; redwood, ∿9200 Btu/lb (Arola, 1976); and southern pines, ∿8950 Btu/lb (Wen, 1974). The gross heat available and lower heating value functions associated with those species are nearly identical to those presented for Douglas-fir. Variations in higher heating value as a function of species occurs, for all practical purposes with a relatively narrow range—(8,000-10,100 Btu/lb) (Corder, 1973). Variations in lower heating value associated with moisture content are, therefore, of more importance since they occur within the range of 8580 to 1650 Btu/lb. This conclusion is supported by Tables 5 and 6 of Chapter 3 in the CORRIM report on energy and chemicals from lignocellulosic materials (Bethel, 1976).

From Fig. 1 it becomes apparent that moisture content is a critical design parameter in establishing wood fueled systems. Moisture significantly increases the heat required for pre-ignition and substantially decreases the net heat release from combustion (Shafizadeh, 1977). It reduces the flame temperature of combustion from a theoretical 3000°F to 1700°F as moisture goes from 0 to ~33% (Levelton, 1978), thus inhibiting efficiency of heat recovery. High moisture fuels (e.g., 60% moisture) can be utilized if that condition is designed into the system, but the penalties in equipment size and boiler efficiency must be paid.

B. *The Cost of Using Wood Fuel to Generate Steam by Direct Combustion*

The process of calculating an opportunity cost curve for chips, planer shavings, sawdust, and bark as fuel involves imputing a $/ton value based on oil saved. As Fig. 1 shows, the mass of wood required to replace a given barrel of oil is a function of moisture content.

Two recent and very complete studies (Levelton, 1978, Bliss & Blake, 1977) have addressed capital and operating costs for wood fueled systems. These studies by B. H. Levelton and Associates, Ltd. and Mitre Corporation address capital and operating costs. Previously Vannelli and Archibold (1976) have estimated costs for larger (150,000 and 400,000 lb/hr) systems, as has Jamison (1978) and Mitre.

Figure 2 is a plot of values for capital cost of wood fueled systems as a function of size, based on values reported in the literature and in interviews.

For the 120,000 lb (150-300 psig)/hr wood fired boilers under consideration here, the Weyerhauser Co. (Jamison, 1978) supplied the estimated 1978 capital and operating cost data presented in Table II. These estimates are consistent with the Vannelli figures and are supported by previously developed U.S. Forest Service estimates. Operating, maintenance, and return on investment estimates have also been made for wood fueled systems based upon assistance by Weyerhauser Co., American Fyr Feeder Engineers (Voss, 1978), and also based upon the following assumptions: (1) a 6% total maintenance charge (Bliss and Blake, 1977), and a 2.5% charge for local taxes and insurance (Ellis, 1978). These estimates are presented in Table III, along with cost/10^3 lb steam Btu. The

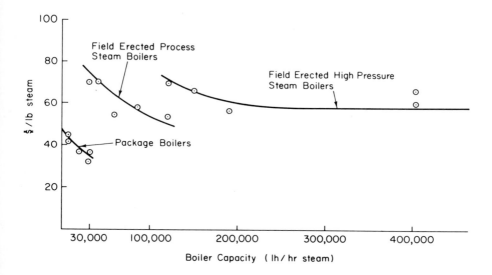

FIGURE 2. Estimated installed capital cost of wood fired boilers by capacity and type (assume 50% M fuel).

estimates of cost/10^3 lb steam are based on the following formula:

$$\frac{.334 \ C_c + C_o + .085 \ C_c}{E_o} = C_E \quad (5)$$

where .334 → capital cost factor as shown in Table II; C_c → capital cost; C_o → operating cost; .085 → 6% of capital cost for maintenance, and 2.5% of capital cost for local and state taxes; E_o → annual energy output, expressed in 10^3 lb steam/yr; and C_E → cost/10^3 lb steam produced.

TABLE III. *The Cost of 120,000 lb/hr (300 psig) Wood Fired Boiler Compared to the Cost of a Comparable Oil Fired Boiler (in $000) (assume 50% moisture fuel).*

Cost category	Wood fired	Oil fired
Capital cost		
Site preparation	106	93
Fuel handling system	939	100
Boiler and related	3692	943
Spares	150	50
Total direct cost	4887	1186
Engineering @ 15%	733	178
Commissioning and conting. @ 15%	843	205
Working capital @ 5%	305	78
Total capital cost	6768	1647
Annual capital cost	2260.5	550.1
Annual operating cost	249.6	124.8
Maintenance @ 6%	406.1	98.8
Taxes and insurance @ 2.5%	169.2	41.2
Fuel cost (oil at $14.23/bbl)	–	3761.0[a]
Total annual cost	3085.3	4575.9

[a] *Assumes 3-shift operation and .8 load factor.*

C. *The Opportunity Cost Curve for Wood Fuels Based on Combustion and Steam Generation*

The basic opportunity cost curve for wood residue as fuel is based upon the total savings associated with not burning oil imputed back to the tons of wood combusted. There are as many capital costs as oil fired boilers ranging from ~$10/lb capacity to ~$20/lb (Kipper, 1978; Axtman, 1978). All of these estimates assume boilers supplying 300 psig saturated steam. For purposes of this analysis, a modified Weyerhauser estimate, as presented in Table III, is assumed since it reflects total installed costs including company expenditures.

Operating and maintenance costs of oil fired systems for the Pacific Northwest have also been estimated by Weyerhauser Co. (Jamison, 1978).

These estimates can be modified as a function of moisture content. Table IV shows the influence of moisture on capital costs and total costs based on the capacity curve of Vannelli and Archibald (1976), and assuming constant operating and maintenance costs.

Conceptually one can then view the opportunity cost curve as being derived from the following formula:

$$\begin{bmatrix} \text{Fuel value} \\ \text{of mill} \\ \text{residues} \end{bmatrix} = \frac{\begin{bmatrix} \text{Annual cost of pur-} \\ \text{chasing and operating} \\ \text{a comparable sized} \\ \text{petroleum fueled} \\ \text{system (\$)} \end{bmatrix} - \begin{bmatrix} \text{Annual cost of pur-} \\ \text{chasing and opera-} \\ \text{ting a wood fueled} \\ \text{system (\$)} \end{bmatrix}}{\begin{bmatrix} \text{Annual wood consumed by wood} \\ \text{fueled system} \end{bmatrix}} \quad (6)$$

Where $V_c \rightarrow$ value of residues for steam raising, $C_c \rightarrow$ annual capital cost, $C_{o,m} \rightarrow$ annual operating and maintenance cost, $C_f \rightarrow$ annual fuel cost, $p \rightarrow$ petroleum system, $w \rightarrow$ comparable wood fueled system, and $WC_{ws} \rightarrow$ annual tons of wood consumed. More precisely, this formula appears as:

$$V_c = \frac{(C_c + C_{o,m} + C_f)_p - C_c + C_{o,m})_w}{WC_{ws}} \quad (7)$$

This approach clearly defines the resultant values as opportunity costs rather than investment analyses for the alternative options, as represented by the second term in the numerator. They will vary on an individual company basis. Further this formula demonstrates the full influence of moisture content on fuel wood value for, as Fig. 1 and Table IV have shown, residue consumption to produce a given amount of energy increases with moisture content. The economic value, therefore, is inversely proportional to moisture content.

The opportunity values of various residues are presented in Table V, based upon the formula (7). It is significant to note the precipitous decline in the economic value of wood residue fuels as a function of moisture content.

TABLE IV. The Influence of Moisture Content on the Economics of a 120,000 lb Boiler

Consideration	Percent Moisture (Gr. Wgt.)		
	50	30	10[a]
Equivalent boiler capacity at 50% moisture (10^3 lb/hr steam)	120	90	75
Installed capital cost (10^3 $)	6768	5175	4837.5
Annual capital cost (10^3 $)	2260.5	1728.5	1615.7
Total annual cost (10^3 $)	3085.3	2533.3	2440.5
Annual savings compared to oil	1490.6	2022.6	2135.4

[a] *The values on 10% moisture are not strictly proportional since the system design includes complete inside storage, more complete fuel preparation methods and pneumatic fuel handling plus extensive fines control systems These systems are largely avoided at 30%M and above.*

TABLE V. Wood Fuel Values Based on Savings From Not Burning Burning Oil

Fuel quality (% moisture)	Tons consumed per year (Green)	Value	
		$/ton (Gr wgt)	$/BDT
10	100,000	21.35	23.73
30	156,500	12.92	18.46
50	214,000	6.95	13.89

[a] *This table is used to calculate the opportunity cost curve for wood as fuel. To estimate the Return-on-Investment, the lost revenues from not selling wood residues as furnish should be used as fuel costs (Jamison, 1978).*

D. The Technological Sensitivity of the Wood Fuel Opportunity Cost Curve

For many applications the cogeneration of steam and electricity is often considered most appropriate for increasing the efficiency of the energy system. Their economic attractiveness can be considered from several vantage points, including the value which they impute to mill residues. Thus it is considered here. For some utilities wood is considered to be a quite viable alternative fuel. Thus it is considered here also.

1. Cogeneration. Cogeneration, the production of electricity and steam in the same unit, is considered particularly appropriate for the pulp and paper industry since pulp mills are large consumers of both forms of energy (Nydick, 1976). It is also useful in large sawmills and structural wood products complexes. There are several approaches to cogeneration including the back pressure turbine or topping cycle depicted in Fig. 3. In this cycle, steam is generated in a high pressure boiler, passed through a steam turbine driving an electricity generator, and the waste steam from the turbine is then employed for process energy purposes. This system is particularly appropriate for wood using industries since it can use solid fuels directly (Gyftopoulos, 1976). Further, most of the energy not used in electricity generation is contained in the enthalpy of the reject steam.

The cogeneration principle is already quite popular in the forest products industry. In the Pacific Northwest some 200 megawatts (MV) of capacity are now contributing to the power grid, and another 200 megawatts of capacity exists, although it remains idle for economic reasons. Thus, some 29% of the technical potential (1400 megawatts) has been developed (Rocket, 1979). Nationwide the Weyerhaeuser Co. has 11 electricity producing cogeneration systems plus 25 direct drive or motive power systems (Methven, 1978). St. Regis employs cogeneration at 12 of its 13 pulp mills (Tillman, 1978). Cogeneration deployment, however, will be increased dramatically in years to come as smaller systems (e.g., 1.5-5.0 MW capacity) become economic.

The technique for estimating the economic value for cogeneration is related to yet distinct from the technique for basic consumption. It considers wood fired steam raising as the base case and imputes a value to the increment of fuel needed to support cogeneration as a function of electricity not purchased. All calculations, then, are based on incremental costs and wood consumption, with the cogeneration

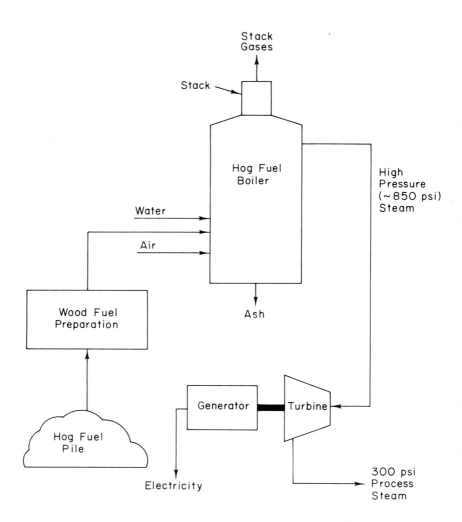

FIGURE 3. Schematic of cogeneration.

increment wood value being rolled into the basic wood value. The two formulae employed are as follows:

$$WV_{cg} = \frac{(E_p \times E_v) - \Delta(C_c + C_{o,m})}{WC_{cg}} \quad (8)$$

$$WV_t = \frac{S_{ws} + [(E_p \times E_v) - \Delta(C_c + C_{o,m})]}{WC_{ws} + WC_{cg}} \quad (9)$$

where:

WV_{cg} = wood value of increment of fuel consumed to support cogeneration

E_p = kilowatts of electricity produced annually

E_v = value of electricity produced (¢/kw)

ΔC_c = change in capital cost required

$\Delta C_{o,m}$ = change in operating cost required

WC_{cg} = annual tons of wood consumed to support cogeneration

WV_t = average value of all wood consumed in cogeneration system

S_{ws} = annual savings defined as $V_s \times WC_{ws}$ in formula (7)

WC_{ws} = wood consumed in basic steam raising system.

These formulae approximate the sequence of events for the investment decision, and make the imputed wood value sensitive to both oil price and electricity price.

Opportunity cost values for wood fuels employed in cogeneration are based upon boiler producing $120,000 lb/hr of 825°F, 875 psig steam generating 5 megawatts (5×10^3 kilowatts) of electricity per hour. The capital, operating and maintenance costs, presented in Table VI have been estimated by Weyerhauser based on a recent installation (Jamison, 1978).

It is significant to note that this incremental investment is moisture insensitive. It is required for the feedwater pretreatment system, the high pressure-high temperature valves, and the turbogenerator. The moisture related

TABLE VI. Incremental Cost of Cogeneration Technology Rated at 5 Megawatts of Power/hr, $120,000 lbs. 200 psig Process Steam (in $000)

Cost item	Cost (10^3 $)
Installed capital cost	1550
Annual capital cost	520
Annual operating cost	50.4
Maintenance	120
Local taxes and insurance	38.8
Total annual cost	729.2

investment costs are in the basic materials handling system and the combuster-boiler system itself.

Given this incremental investment, one can then calculate the electricity cost which at least equates cogeneration with basic wood combustion for low pressure steam production and then calculate the imputed value of wood with electricity costing 2.81¢/KWH (average cost of U.S. industrial power) and with electricity costing 3.5¢/KWH (the average cost of new generation). These values are presented in Table VII.

2. Electricity Generation. Electricity generation in full condensing turbines fueled by wood is being considered and developed by numerous utilities. Burlington, Vermont is now home of a 5 MW power plant and Green Mountain Power is in the process of deploying a 50 MW unit. Washington Water Power has announced plans to build a 40 MW unit. Consumers Power in Michigan has announced a 25 MW unit.

The influence of moisture content on fuel value can be seen most easily from the plant heat rate. For 50% M fuels, it approaches 20,000 Btu/KWH. For 30% M fuels, it is ∼15,000 Btu/KWH and for 10% M fuels, it may be as low as 13,000 Btu/KWH. The heat rate for a modern coal fired plant is ∼9500 Btu/KWH. Heat rate is also influenced by plant size. This is a particular problem since 50 MW is generally considered to be the largest practical size for a wood fueled unit. For purposes of estimation here, the 50 MW unit is used.

TABLE VII. Imputed Wood Fuel Values Associated with Cogeneration

	Percent Moisture		
Consideration	50	30	10
Annual wood consumed (10^3 ton)	237.8	173.8	112.5
Breakeven price of electricity (¢/KWH)	2.23	2.38	2.49
Imputed total wood value @ 2.81¢/KWH ($/Green ton)	7.93	13.91	22.49
Imputed total wood value @ 2.81¢/KWH ($/BDT)	15.86	19.87	24.99
Imputed total wood value @ 3.5¢/KWH ($/Green ton)	9.09	15.50	24.94
Imputed total wood value @3.5¢/KWH ($/BDT)	18.18	22.14	27.72

Because the product of these systems is a marketable form of energy, opportunity value estimation can be performed by a modification of the incremental formula (8) of cogeneration:

$$WV_e = \frac{(E_p \times E_v) - (C_c + C_{o,m})}{WC_c} \tag{10}$$

where WV_e → wood value from electricity production, E_p → annual electricity production, E_v → electricity value, C_c → annual capital cost, $C_{o,m}$ → annual capital and operating cost, and WC_e → wood consumed annually to produce electricity. In order to employ this formula, Table VIII presents base case costs for electricity production assuming a 50 MW plant and an annual capital cost of 15.85% as shown in Table II. Costs were taken from Bliss and Blake (1977).

TABLE VIII. Base Case Costs for Electricity Production (50 MW plant)

Cost item	Cost ($000)
Direct capital cost	37,800[a]
Engineering	5,700
Contingency and commissioning	6,500
Working capital	2,500
Total installed cost	52,500
Annual capital cost @ 15.85%	8,300
Annual operating costs	3,900
Total annual cost	12,200
Total electricity produced (8000 hrs)	440,000,000 KW
Cost without fuel	27.7 mills/KWH

[a] Capital costs have been held constant here, consistent with values reported in the literature. The heat rate approach effectively makes the compensation for water content.

The conversion of base costs into economic values is shown in Table IX. It is shown at 3.5¢/KWH for national marginal power costs, and at 5.0¢/KWH to reflect costs in northern New England.

3. Technological Conclusions. Based on these data for direct combustion, cogeneration, and electricity production by condensing turbines, a series of value estimating equations may be developed. These are as follows, assuming a constant oil price of $14.25/bbl. For electricity generation by utilities, formula (13) assumes a value of 3.5¢/KWH.

$$FV_c = 24.54 - 0.36M \tag{11}$$

$$FV_{cg} = 19.12 + 2.51E - 0.38M \tag{12}$$

$$FV_e = 9.75 - 0.135M \tag{13}$$

where $FV_c \to$ fuel value in combustion systems, $FV_{cg} \to$ fuel value in cogeneration systems, $FV_e \to$ fuel value in full

TABLE IX. The Influence of Electricity Price and Moisture Content on Wood Value in Electricity Generation

Item	Fuel moisture content		
	50%	30%	10%
Total annual wood consumed (10^3 tons)	1,035[a]	520[b]	375[c]
Annual product value @35 mills/KWH ($000)	15,400	15,400	15,400
Wood value ($/ton) @35 mills/KWH	3.10	6.15	8.55
Wood value ($/BDT) @35 mills/KWH	6.20	8.20	9.50
Annual product value @50 mills/KWH ($000)	22,000	22,000	22,000
Wood value ($/ton) @50 mills/KWH	9.45	15.00	20.80
Wood value ($/BDT) @50 mills/KWH	18.90	25.10	29.05

[a] Heat rate = 20,000 Btu/KWH.
[b] Heat rate = 15,000 Btu/KWH.
[c] Heat rate = 13,000 Btu/KWH.

condensing turbines, M → percent moisture (green weight basis), and E → cost of electricity in ¢/KWH. All economic values are presented on a green ton basis.

Of the three systems cogeneration imputes the highest value to wood fuels generally, and is also most sensitive to moisture content. Full condensing turbines generally impute the lowest value to wood fuels and are least sensitive to moisture content.

V. THE VALUE OF WOOD RESIDUES IN THE MANUFACTURE OF MATERIALS

The energy users within and outside of the wood products industry must compete with materials users of residues such as pulp mills and particleboard plants. It is this competition which creates the need for opportunity value estimation. Wood residues have become a primary source of raw materials for the pulp and paper industry, and are the sole source of raw materials for the particleboard industry. Chips and, to a lesser extent sawdust, are used in pulp manufacture while sawdust and shavings are furnish for particleboard and hardboard plants. Competition for these chips, planer shavings, and sawdust for the most part has favored the materials markets. Residues delivered to Weyerhauser plants currently are valued as follows: pulp chips, $30/BDT; planer shavings, $7/BDT; sawdust, $6/BDT; and bark (as fuels), $4/BDT (Jamison, 1978).

The influence of moisture on residues sold to pulp and particleboard mills is unrelated to its economic value. Above the fiber saturation point ($\sim$25% M) moisture does not influence value. Residues below fiber saturation may command a somewhat lower price as a result of incomplete cooking of dry materials in the pulp digester. This causes economic penalties in the pulping process. Thus for all practical purposes, the curves are linear, a function of the fiber content without added penalties.

The value of residue as materials, to the seller, is a function of distance from the customer. In 1971, Hyde and Corder demonstrated that transportation costs are weight limited rather than volume limited, and linear (Hyde and Corder, 1971); facts confirmed by Jamison (1978). Estimates of transportation costs made by the Weyerhuaser Co. are presented in Table V. The influence of transportation costs on the economic value of wood residues also is presented in Table VIII. In effect Table X defines the opportunity cost curve for wood residue materials in the Pacific Northwest assuming typical moisture contents of $\sim$33% M (dwb) for chips and sawdust, and $\sim$10% M for planer shavings.

Table X is particularly interesting for it shows that after the first 10 miles, the truck transport function for pulp chips is as follows:

$$SV_{pc,t} = 28.39 - 0.12D \qquad (12)$$

where SV → sales value and D → distance in miles. For planer shavings the function is:

TABLE X. Transportation Costs and Their Impact on the Sales Value of Pulp Chips and Planer Shavings (in $/BDT)

	Transportation cost		Sales value of residues[a]	
Mileage	Truck	Rail	Pulp chips	Planer shavings[b]
0	–	–	30.00	7.00
10	2.77	7.97	27.23	4.70
20	4.00	7.97	26.00	3.68
30	5.24	7.97	24.76	2.65
40	6.47	7.97	23.53	1.63
50	7.70	7.97	22.23	0.61
60	8.93	8.32	21.68	0.09
70	10.16	8.65	21.35	
80	11.40	8.97	21.03	
90	12.63	9.42	20.58	
100	13.86	9.88	20.12	
150		11.60	18.40	
200		13.30	16.70	
250		14.90	15.10	
300		16.70	13.30	
400		19.02	10.98	
500		21.31	8.69	
600		23.36	6.64	

[a] Gross delivered value – transport costs.
[b] Shavings transportation costs 83% of pulp chip transport.
Source: (Jamison, 1978).

$$SV_{ps,t} = 5.72 - 0.10D \tag{13}$$

At 60 miles rail transportation becomes more economic, with a cost function for pulp chips (above that distance) of:

$$SV_{pc,r} = 22.91 - 0.029D \tag{14}$$

These three functions become the price predictors analagous to the ones previously presented. Most interesting is the dry fuel transport function:

$$SV_{df} = 22.20 - 0.12D \tag{15}$$

Theoretically one could transport shavings for fuel 125 miles before their furnish value would equal their fuel value.

There is, of course, an alternative method of treating or considering residues, particularly bark: that is as landfill. This approach has a negative value which is site specific depending on: (1) soil conditions; (2) rainfall; (3) proximity to a lake, river, or other water body; (4) population density of industry environs; (5) local regulations; and (6) remaining landfill capacity. Costs/ton or costs/yd^3 can vary wildly, and averages have little meaning. Further, as one looks to the future, such practices will be made increasingly obsolete by substantial oil price increases and environmental regulations. Thus they will receive no further attention.

VI. OPPORTUNITY COST COMPARISONS

The comparative questions posed can now be addressed. Restated they are: (1) at what moisture contents are residues more valuable as fuel than furish, (2) how does oil price affect the relative values of residues as fuel and furnish, and (3) what is the influence of transportation distance between residue producer and consumer on the relative value of such materials. In answering the first two questions, no transportation is assumed. In answering the third question, it is assumed that wood residues would be consumed on-site as fuel, but transported if used as furnish. In answering the second and third questions, typical moisture contents are assumed.

A. General Economic Value Curves

Figure 4 compares the economic value of residues as fuel for chips (at 20-60% M) shavings (2.5-20% M), and sawdust (2.5-60% M); the moisture ranges selected are those which occur in industrial practice. Significant to note are the following: (1) on an oil saved basis, pulp chips always remain more valuable as furnish; (2) planer shavings always are slightly more valuable as fuel; and (3) sawdust, with its wider moisture range, may be more useful in either application depending upon whether it is produced (wet) in a lumber mill or (dry) in a cut stock or furniture plant. Its use is also technology sensitive. In Fig. 4 electricity prices are arbitrarily held at 3.5¢/KWH for graphic purposes when cogeneration is considered.

Figure 4 is the central figure. Thus it should be cautioned that the residue prices shown in the figure reflect a currently depressed market. For example, pulp chips exported to Japan were valued at $49.17/BDT (shipped from Washington) in 1977, a value which fell to $41.31 by the first quarter of 1978 (Ruderman, 1978). A return to better market conditions would raise the value of pulp chips and probably also sawdust—swinging the balance of that curve. It should also be observed that electricity from full condensing turbines has not been shown. As has been noted previously, such systems impute low values to wood fuels— opportunity values which will make utility use of wood fuels subject to special case rather than general case conditions. Process steam and cogeneration systems are more universal.

Despite the present residue furnish market conditions, Fig. 4 demonstrates the economic reasons for Moore's survey findings: that coarse residues are virtually always sold to pulp mills, that sawdust is most commonly used as fuel, and that shavings are far more commonly used as fuel than had been anticipated (Moore, 1976). It also confirms the high rate of return and short pay-off period which Furman and Desmon show for the Crown-Zellerbach, Omak, Washington, veneer drying system fueled with dry veneer clippings, planer shavings, and cut stock residues (Furman and Desmon, 1976).

Because Fig. 4 is central, one additional caution merits comment—that this figure is not necessarily an argument for predrying and/or densification. While such processes are advocated by Vannelli and Archibald (1976), and by other authors, conclusive evidence of their cost effectiveness has not been shown. Problems of reliability, low grade oxidation, air and water pollution, and thermal efficiency have been cited as objections to drying. Thus Fig. 4 should be viewed as a series of value curves without regard to additional processing.

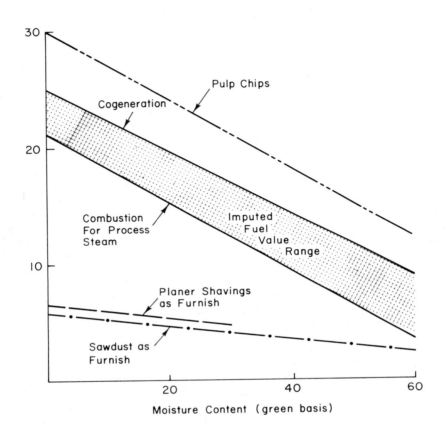

FIGURE 4. The relative value of residues as fuel and furnish.

B. The Influence of Rising Energy Prices on Wood Fuel Values

Rising oil and electricity prices have a decided influence on the value of wood residues and the appropriate market for those residues. The issues are somewhat complex, but can be dissected as follows: (1) changes in oil price relative to electricity price, and (2) a uniform rise in oil and electricity prices.

The issue of oil prices is particularly important in light of the dramatic rises in price resulting from the Iranian revolution. While OPEC prices of $14.25 were in place, numerous countries sold oil at levels considerably higher. Spot prices exceeded $20/bbl. ($3.45/10^6 Btu). The world oil surplus, which served to dampen price rises, vanished. Thus price hikes can only escalate in the months and years to come.

1. The Changes in Relative Oil and Electricity Prices. The change in electricity price relative to oil price does not determine at what point residues are more valuable as fuel or as furnish. Rather changes at this level determine the economics of process steam raising vs. cogeneration. Formulae (7) and (8) can be used to set a current relationship which is $14.23/bbl. oil and 2.23¢/KWH for 50% M fuels. If electricity price falls below 2.23¢/KWH relative to the current oil price, cogeneration becomes economic. If oil prices rise relative to the 2.23¢/KWH, cogeneration also becomes uneconomic. Thus, as oil prices rise, electricity prices must keep pace. At $20/bbl. of oil, electricity must cost at least 3.0¢/KWH for cogeneration to command wood resources away from a system of raising process steam with hog fuel and buying power from local utilities.

2. The Change of Energy Prices Relative to All Other Prices. As the price formulae (11, 12, 13) showed, all economic values are sensitive to moisture content. If oil and electricity prices rise, how will wood fuel values respond? It can be shown that the fuels will increase in value at an even pace, with a change in intercept value being the only real difference in price functions. Table XI, using direct combustion for steam raising only as the technology, shows the influence of oil price on imputed value of wood residue in $/BDT. It shows that relative to a low value of pulp chips of $30/BDT, an oil price 1.5 × current prices will establish fuel-pulp chip competition. The high pulp chip price reported of $49/BDT requires a doubling in oil price with no other changes for competition. Of course a rise in oil prices will cause an increase in pulp chip prices. Thus

TABLE XI. The Influence of Oil Price on Imputed Value of Wood Residues (Moisture contents are green wgt basis)

Oil price ($/bbl)	Imputed fuel value ($/BDT)		
	50%M fuel[a]	30%M fuel[b]	10%M fuel[c]
14.23	13.89	18.46	23.73
21.35	31.42	35.63	40.16
28.46	48.94	52.79	58.96
35.58	66.47	69.98	76.16
42.69	83.98	87.14	93.76

[a] $WV_{50} = 2.463P - 21.162$
[b] $WV_{30} = 2.413P - 15.872$
[c] $WV_{10} = 2.475P - 11.898$

$WV \rightarrow$ wood value, $/BDT/
$P \rightarrow$ oil price, $/BBL.

real competition may be anticipated only in the >$30/bbl. oil price region.

In order to assemble the influences of energy prices in a convenient form, Table XII has been constructed. Opportunity value equations have been developed for combustion to raise low pressure steam for cogeneration and for electricity generation by full condensing systems owned by utilities. From these equations one can easily see the influence of both oil and electricity price, and their need to march in lockstep if cogeneration is to remain economic.

C. The Influence of Transportation Costs on Residue Values

It has been shown that, at current residue furnish values, planer shavings are already somewhat more valuable as fuel than as furnish as is most sawdust. Does transportation influence the tradeoff between using coarse residues for pulp or energy? Table XI shows the fuel value of 30% M chips to be $18.46/BDT if they are not transported. At distances of less than 150 miles, the selling price of pulp chips (delivered value-transportation cost) exceeds $18.46/BDT. Beyond that transportation distance there is such a tradeoff.

TABLE XII. Opportunity Value Functions by Moisture Content and Technology for Various Energy Prices

Moisture content (% G.W.)	Technology	Function	
50	Combustion-steam raising	$V_{c,50}$	$= 1.23F - 10.53$
	Cogeneration	$V_{eg,50}$	$= 1.08F + 1.64E - 12.27$
	Electricity (full condensing)	$V_{e,50}$	$= 4.45E - 12.34$
30	Combustion	$V_{c,30}$	$= 1.69F - 11.11$
	Cogeneration	$V_{cg,30}$	$= 1.52F + 2.21E - 14.06$
	Electricity	$V_{e,30}$	$= 7.69E - 23.46$
10	Combustion	$V_{c,10}$	$= 2.64F - 16.26$
	Cogeneration	$V_{cg,10}$	$= 2.35F + 3.56E - 20.99$
	Electricity	$V_{e,10}$	$= 10.66E - 32.53$

Definitions:

V_c = imputed wood value, combustion for steam system;
V_{cg} = imputed wood value, cogeneration system;
V_e = imputed wood value, electricity generation by full condensing turbine;
V_m = imputed wood value from methanol production;
F = imported oil price ($/bbl); and
E = marginal cost of electricity generation (¢/KWH).

VII. CONCLUSION

Three moisture questions are related to the economic value of using wood residues as fuel: (1) the energy content of the fuel; (2) the cost of building and operating a wood-fired system, and the resulting savings from not using oil; and (3) the cost of moving residues from place to place. Resulting from a consideration of those factors are series of opportunity cost curves which estimate how those moisture questions influence the relative values of using residues for fuel or furnish. Given the current market prices for residues as reflected in the Weyerhauser data, the following conclusions can be reached.

(1) planer shavings and sawdust produced are now more valuable as fuel than as furnish;

(2) technology, specifically cogeneration, plays a key role in imputing values to wood fuel;

(3) a doubling of oil prices relative to all other values would direct all mill residues save pulp chips to energy markets, initiate some competition for those coarse residues, and could introduce whole tree chips into the energy marketplace in significant quantities; and

(4) because of relatively high oil prices and depressed residue values, transportation does not affect the tradeoff point, save pulp chips at >150 miles.

Moisture then is one critical variable in directly residues to their point of maximum economic value. As one looks to the future, this technique of imputing values back to wood residues and/or harvested materials can be employed as a method for comparing the relative economic merits of such technologies as methanol production and high Btu ($\sim$1000 Btu/scf) gasification; or systems development in such areas as silvicultural energy farming.

ACKNOWLEDGMENT

I would like to acknowledge the assistance of Robert Jamison, Director, Energy Management Group, Weyerhauser Co. who supplied considerable information, advice, and critical comment.

REFERENCES

Arola, Roger (1976). "Wood Fuels--How Do They Stack Up. Energy and the Wood Products Industry." Proceedings of the FPRS Meeting, Atlanta, November 15-17, 1976. Forest Products Research Society, Madison, Wisconsin.

Axtman, William (1978). American Boiler Manufacturers Assn., Personal communication.

Bethel, J. S. et al. (1976). "The Potential of Lignocellulosic Material for the Production of Chemicals, Fuels, and Energy." National Academy of Sciences (Committee on Renewable Resources for Industrial Materials), Washington, D.C.

Bliss, C. and Blake, D. O. (1977). "Silvicultural Biomass Farms, Vol. V: Conversion Processes and Costs." The Mitre Corporation [for ERDA, Contract E(49-18)-2081].

Bogot, Ralph (1976). "Wood as a Total Energy Source." Presented at the International Woodworking, Machinery and Furniture Supply Fair, Louisville, Kentucky, September 18-22.

Christensen, G. W. (1975). "Wood Residue Sources, Uses, and Trends. Wood Residue as an Energy Resource." Proceedings of the FPRS Meeting, Denver, September 3-5. Forest Products Research Society, Madison, Wisconsin.

Corder, S. E. (1973). "Wood and Bark as Fuel." Forest Research Laboratory, Oregon State University (Research Bulletin No. 14).

Duke, J. M. (1978). Report to the Department of Energy on Energy Conservation: Progress of the Pulp, Paper and Paperboard Industry in 1977. Raw Materials and Energy Division, American Paper Institute, New York.

Ellis, T. H. (1978). Economic Analysis of Wood- or Bark-Fired Systems." U.S. Forest Service (General Technical Report FPL 16), Madison, Wisconsin.

Furman, L. H. and Desmon, L. G. (1976). Wood residue for veneer drying: A case history, *Forest Products Journal*, September.

Furman, Lloyd H. (1978). Crown-Zellerbach, Omak, Washington. Personal interview.

Grantham, J. B., Estep, E. M., Pierovich, J. M., Takow, H., and Adams, Thomas C. (1974). "Energy and Raw Material Potentials of Wood Residue in the Pacific Coast States - A Summary of a Feasibility Investigation." Pacific Northwest Forest and Ranger Experiment Station, Portland, Oregon.

Gyftopoulos, Elias, et al. (1976). "A Study of Improved Fuel Effectiveness in the Iron and Steel and Paper and Pulp Industries." Report of Thermo Electron Corp. to the National Science Foundation.

Hall, E. H. et al. (1976). "Comparison of Fossil and Wood Fuels." Battelle Columbus Laboratories (for the Environmental Protection Agency), Columbus, Ohio.

Ince, Peter J. (1977). "Estimating Effective Heating Value of Wood or Bark Fuels at Various Moisture Contents." USDA Forest Service (General Technical Report FPL 13), Madison, Wisconsin.

Jamison, Robert L. (1977). "Trees as a Renewable Resource. Clean Fuels from Biomass and Wastes." Proceedings of a conference, January 25-28, at Orlando, Florida. Institute of Gas Technology, Chicago, Illinois.

Jamison, Robert L. (1978). Weyerhauser Co. Personal communications.

Levelton, B. H. and Associates, Ltd. (1978). "An Evaluation of Wood Waste Energy Conversion Systems." Environment Canada, Vancouver, B.C.

Methven, Norbert E. (1978). "Existing Cogeneration Facilities." Presented at the Pacific Northwest Bioconversion Workshop, Portland, Oregon, October 24-26.

Moore, William E. (1976). Wood residue energy conversion systems market, *Forest Products Journal,* March.

Nydick, S. E. et al. (1976). "A Study of Inplant Electrical Power Generation in the Chemical, Petroleum Refining and Paper and Pulp Industries." Report of Thermo Electron Corp. to the Federal Energy Administration.

Riley, John G. (1976). "Development of a Small Institutional Heating Plant to Utilize Forest Residue Fuels." Presented at the 1976 Annual Meeting, North Atlantic Region, American Society of Agricultural Engineers, New Brunswick, New Jersey, August 15-18.

Rocket Research Co. (1979). Industrial Electrical Cogeneration Potential in the Bonneville Power Administration Service Area. Bonneville Power Administration, Portland, Oregon, January 19.

Schlesinger, M. D., Sanner, W. S., and Wolfson, D. E. (1972). "Pyrolysis of Waste Materials from Urban and Rural Sources." Proceedings, Third Mineral Waste Utilization Symposium, March 14-16.

Shafizadeh, Fred, and DeGroot, William F. (1976). Combustion characteristics of cellulosic fuels, in "Thermal Uses and Properties of Carbohydrates and Lignins" (Fred Shafizadeh, Kyosti V. Sarkanen, and David A. Tillman, eds.). Academis Press, New York.

Shafizadeh, Fred, and DeGroot, William F. (1977). Thermal analysis of forest fuels, in "Fuels and Energy from Renewable Resources" (David A. Tillman, Kyosti V. Sarkanen, and Larry L. Anderson, eds.). Academic Press, New York.

Tillman, David A. (1977). Combustible renewable resources, *CHEMTECH*, October.

Tillman, David A. (1978). "Wood as an Energy Resource." Academic Press, New York.

U. S. Forest Service (1976). "The Feasibility of Utilizing Forest Residues for Energy and Chemicals." Report to the National Science Foundation. Forest Service, USDA, Washington, D.C.

Vannelli, L. S. and Archibald, W. B. (1976). "Economics of Hog Fuel Drying. Energy and the Wood Products Industry." Proceedings of the FPRS Meeting, Atlanta, Georgia, November 15-17.

Voss, George (1978). American Fyr Feeder Engineers. Personal communication.

Wen, C. Y. et al. (1974). Production of low Btu gas involving coal pyrolysis and gasification, in "Coal Gasification" (Lester G. Massey, ed.). American Chemical Society Press, Washington, D.C.

PYROLYSIS OF WOOD RESIDUES WITH A VERTICAL BED REACTOR

J. A. Knight

Engineering Experiment Station
Georgia Institute of Technology
Atlanta, Georgia

I.	INTRODUCTION	88
II.	BACKGROUND INFORMATION	89
III.	DEVELOPMENT OF GEORGIA TECH/TECH-AIR PYROLYSIS SYSTEM	92
	A. Pilot Plant Development at Georgia Tech	92
	B. Development of Tech-Air Field Demonstration Units	98
IV.	GEORGIA TECH/TECH-AIR PYROLYSIS PROCESS	101
	A. Description	101
	B. Mass and Energy	104
	C. Comparison with Badger-Stafford Process	106
	D. Pyrolysis versus Gasification	107
V.	PRODUCTS AND UTILIZATION	109
	A. Gases	109
	B. Pyrolysis Oils	110
	C. Charcoal	111
	D. Combustion Tests	112
VI.	MOBILE PYROLYSIS SYSTEM	113
VII.	SUMMARY	114

I. INTRODUCTION

Pyrolysis of wood and other lignocellulosic materials in the absence of air or very limited amounts of air produces charcoal, condensible organic liquids, water, and noncondensible gases. Pyrolysis has been used since the earliest times of civilization to produce charcoal from wood and related materials for use as a domestic and an industrial fuel and for smelting ores. In the 19th Century and into the early 20th Century, charcoal was produced on a large scale for use in smelting iron ore in the more industrialized countries. In time, with the depletion of the forests, coke from coal was developed as a replacement for charcoal. Brazil which does not have resources of metallurgical grade coal, still uses charcoal extensively in the steel industry.

If the off-gases from the pyrolysis of wood are collected in some manner, then the process is termed wood distillation. It is of interest that the Egyptians practiced wood distillation by recovering tars and pyroligneous acid from wood pyrolysis for use in their embalming processes. Wood distillation was a profitable industry in the late 19th Century and early 20th Century. In addition to charcoal, wood distillation produced soluble tar, pitch, creosote oil, chemicals and noncondensible gases which were used as a fuel in the plant. With the advent of the petrochemical industry, wood distillation was abandoned in the U.S. Pyrolysis or carbonization of wood and wood residues is still practiced, however, in the U.S. to produce charcoal for the briquette market and for the production of activated carbon.

In more recent times, pyrolysis has received considerable attention as a method for disposal of wastes in an environmentally acceptable manner with practical resource recovery at the same time. Initially, the major emphasis was in developing pyrolysis processes for municipal refuse. The Garrett flash pyrolysis process and the Union Carbide Purox process (use of pure oxygen) are two examples of systems developed for this purpose. The initial focus in developing the Georgia Tech/Tech-Air (GT/T-A) pyrolysis system however, was to process agricultural residues, specifically peanut hulls. The objective of the initial work was to develop a method to dispose of the peanut hulls in a nonpolluting manner with recovery of the charcoal for producing briquettes. More recently, the emphasis has been on the processing of forestry wastes for the recovery of maximum amount of energy from the wastes as charcoal, oil and combustible gases.

The work in pyrolysis at Georgia Tech was initiated in the late 1960s with the objective of developing a low cost, nonpolluting incineration process for peanut hulls. The decision was made after some preliminary experiments to investigate pyrolysis of the hulls with recovery of the charcoal for the home briquette market. The results from the initial work with a hand operated kiln were encouraging and promising, and the decision was made to build a 4,000 pounds per hour pilot plant to investigate the pyrolysis of hulls on a larger scale.

Two parallel efforts took place in the ensuing years to bring the Georgia Tech/Tech-Air pyrolysis system to its present state of development. On the Georgia Tech campus, four pyrolysis pilot plants have been designed, fabricated and tested with a wide variety of feed materials including agricultural, forestry and municipal residues and wastes. Two of these pilot plants are currently in use for testing and research and development programs for improvement of the process. The other effort started with the licensing of the pyrolysis technology by the Tech-Air Corporation in 1970. Based on the results and data from the work with the first pilot plant (4,000 lbs/hr) with peanut hulls, Tech-Air made the decision to design, construct and install a 50 dry ton per day field test and demonstration unit at a peanut shelling plant in south Georgia. This unit was operated for about one year processing hulls from the plant.

The successful results with this unit led Tech-Air in 1972 to design and build a field test pyrolysis facility to process forestry wastes to produce charcoal, oil and combustible gas. This plant, located at a lumber mill in Cordele, Georgia, was operated for a number of years and has proven the process to be technically feasible with better than 90% on-time of rated capacity. In 1975, the American Can Company acquired the Tech-Air Corporation as a wholly owned subsidiary. The interests of the American Can Company centered around the potential use of a pyrolysis system, through their Americology Division, in a total resource recovery facility for municipal refuse and the potential for pyrolysis as a conversion process to produce conventional energy forms from biomass.

II. BACKGROUND INFORMATION

Charcoal, as noted above, has been used by mankind since the earliest times of civilization as a fuel and in certain processes, such as smelting of ores. As civilization progressed, man learned to recover tars, pitch and other materials from the off-gases (smoke) from the incomplete combustion of wood. This latter development lead to the wood distillation industry which was practiced extensively in the industrialized countries in the 1800s and well into the 1900s. Since the subject of this paper is very closely related to the methods used in the early 1900s for production of charcoal and for the wood distillation industry, some background information on these methods is necessary to set the proper prospective of the present day efforts in pyrolysis with the past technology. The present paper could be entitled "Wood Distillation in a Vertical Bed Reactor."

The earliest means used by man for making charcoal in quantity is the technique of covering burning wood with dirt, sod or turf or firing the wood in a hole in the ground. The process is very simple and involves starting a fire with the wood and then reducing the amount of air to just enough to complete the carbonization. Primitive methods similar to these are still used in many sections of the world. From these primitive techniques kilns of many types with improvement of control were developed for producing charcoal with some improvement in efficiency and quality. Two examples of these types of kilns that are in use today are the Missouri kiln (Jarvis, 1959) and the Beehive oven (Simmons, 1964).

Missouri kilns are rectangular buildings of varying dimensions with approximately eight foot walls and an arched roof. The capacities of the kilns vary from 5 to 100 cords. The kilns are provided with controlled air inlet holes in the side and end walls and chimneys along the side walls. The larger ones are provided with doors wide enough to allow trucks to be driven into the kiln. Missouri kilns are presently used to produce charcoal for briquette manufacture in the U.S. Beehive kilns are circular and have walls varying in height of 8 to 20 feet. The kilns are provided with controlled air inlets and exits for the gaseous phase. In Brazil, Beehive ovens are used extensively to produce charcoal in large quantities from Eucalyptus for use in the blast furnace operations of the steel industry. These methods are batch processes in which the wood is tacked in the kiln or oven and ignited, and then the amount of air introduced is controlled so that the wood is carbonized. After cooling, the charcoal is removed.

The addition of a simple condenser to the charcoal kilns provided a means for recovery of some of the condensible material from the off-gases and represented the beginning of the wood distillation industry. The efficiency of these operations was very low, but the method was used extensively when charcoal was produced in large quantities for iron furnaces. The next step in the development of the wood distillation industry was the use of externally heated fixed retorts. This was followed by the development of internally gas-heated retorts. The retort method is a batch process, and the next logical step in both charcoal production and wood distillation was the development of continuous processes.

Many retort designs were proposed and utilized for the continuous production of charcoal, and some of these are discussed by Simmons (1964). These methods were really semi-continuous in that after a retort is loaded and the process started, the removal of charcoal and the addition of wood was carried out on a periodic basis. The wood used in these retorts was of relatively large size, varying in length from 12 to 16 inches and from one to three inches in the other dimensions. The addition of condensers to recover the condensible off-gases converted the system to a wood distillation operation.

A successfully operated wood distillation plant which utilized wood waste has been described by Nelson (1930). This plant was operated by the Ford Motor Company at Iron Mountain, Michigan, in the 1930s to process approximately 400 tons of scrap wood per day. A more detailed description of this operation is presented at this point as a comparison will be made with it after the Georgia Tech/Tech-Air process is described later. The output of this plant was charcoal, noncondensible gas, chemicals (methanol, acetone, ethyl acetate), soluble tar, pitch and creosote oil. Two types of retorts were used to pyrolyze the wood waste. For feed material varying in size from that a match up to 8 × 2 × 3/4 inches, three Badger-Stafford retorts used wood dried to approximately 0.5% moisture. They were operated by utilizing the heat of the exothermic pyrolytic reaction of the wood near the center of the retort. The hot gases from this zone heated the descending wood up to the temperature necessary to initiate the reaction. The average temperature in the central zone was 950°F. The retorts were operated continuously on a two-week cycle. It was necessary at the end of a two-week run to take a retort out of operation and burn out the tar which had accumulated in it. Two Seaman retorts were used to process the sawdust and shavings. The Seaman retorts were slightly inclined rotating kilns about 30 feet long and three feet in diameter and was externally fired. The off-gases from both types of retorts were passed into condensers to recover the tars and pyroligneous acids which were processed for recovery of chemicals, pitch, tars, and oils. The noncondensible gases had a heating value of 290 Btu/scf, and were essentially free of nitrogen. No process air was added to the pyrolysis retorts.

III. DEVELOPMENT OF GEORGIA TECH/TECH-AIR PYROLYSIS SYSTEM

A. Pilot Plant Development at Georgia Tech

In the late 1960s the disposal of peanut hulls in an acceptable manner presented a very real problem to the peanut shelling plants in south Georgia. Open-burning methods and the use of "tepee" burners for the disposal of the hulls were unacceptable because of the environmental problems created by the emissions from these burning operations. The Engineering Experiment Station (EES) at Georgia Tech was assigned the task of developing an incineration method to dispose of peanut hulls in an environmentally acceptable manner.

After some initial investigation and evaluation of the inherent combustion characteristics of peanut hulls, it was decided that incineration of the hulls was not the best approach to disposal. Pyrolysis was a better alternative. The advantages are that a pyrolysis process produces charcoal, oil, and gas, which are three useful fuels. The wastes can then be thought of as resources and the disposal cost can be reduced or possibly eliminated. The charcoal has value as a fuel and can be used in the same manner as coal. It can also be converted to activated carbon, a higher value product. The oil can be used as a fuel or as a chemical feedstock. The noncondensible gases, which have a low heating value per cubic foot, are combustible and can be used on-site as a fuel.

A pyrolysis process developed for peanut hulls would be applicable to other wastes such as forestry and municipal wastes, and therefore, a successful process should find wide utilization as a method for converting wastes into useful materials and fuels. The first pyrolysis experiments on a pilot plant scale were conducted in 1968 in a retort approximately five feet high which had an electric starter, an air inlet device, and a moveable grate. The retort was loaded with peanut hulls, and after initiation of the reaction with the electric starter, the reaction was controlled by the air input rate. Periodically, additional peanut hulls were added and charcoal removed. In the initial experiments, no attempt was made to condense any oil from the off-gases. After processing several hundreds of pounds of peanut hulls, enough data were obtained to show that peanut hulls could be pyrolyzed continuously with a self-sustained reaction in a vertical bed reactor, and charcoal could be produced in a reasonable yield. The results showed enough promise that a decision was made to design and fabricate a pilot plant for a continuous large scale pyrolysis of peanut hulls.

The first continuous pilot plant was designed, fabricated and assembled in 1970 (Knight et al., 1974). It was designed for short life and minimum cost. This unit was approximately 11 feet in height and was designed to process 4,000 pounds of feed material per hour. The pyrolysis reaction in this pilot plant was initiated by means of a natural gas burner and sustained by bleeding a small amount of air into the reactor. The reactor was a vertical bed and gravity fed with the pyrolysis gases passing up through the descending bed. The charcoal was removed at the base of the reactor with a horizontal screw. The off-gases were treated as potentially explosive in these tests; consequently, a system of four flare stacks was constructed to burn the gases in an unconfined, diffusion-controlled flame. Experience with these gases showed that they could be burned safely and easily by premixing them with air and igniting them in a conventional fashion. The pilot plant was operated successfully for about one year with peanut hulls. The initial objective of the test program with peanut hulls was to determine if the vertical bed concept for pyrolysis could be scaled up successfully. If the results of the initial test program showed that the scaled-up pilot plant could be operated successfully, then the next objective was to obtain the necessary data for the design of a 50 dry ton per day field demonstration unit. Peanut hulls were obtained from shelling plants in south Georgia. In the shelling operation, the peanuts are dried before being shelled, and consequently, they have a low percentage of moisture and do not require any drying prior to being pyrolyzed. The test program started with low temperature operation, and on succeeding tests, the temperature in the reactor was increased with eventual damage to the interior surfaces of the reactor. It was not unexpected that some damage would occur because the initial design criteria was based on a low cost unit. The operation and results of this test program with peanut hulls were successful. The tests demonstrated that a vertical bed, gravity fed reactor could be used to pyrolyze peanut hulls in a self-sustained reaction mode to produce charcoal and off-gases. The importance of proper air distribution was a significant result of these tests. Overall the results were very encouraging, and a decision was made by the EES to design and construct a smaller pyrolysis pilot plant for testing with a variety of feed materials.

The pyrolysis process was licensed to the Tech-Air Corporation in 1970 for commercialization. Based on the results of the test program, the decision was made by Tech-Air to design a 50 dry ton per day pyrolysis reactor for a field demonstration program at a peanut shelling plant. The operation and results of the field demonstration units will be discussed in Section III, B.

The second pilot plant was designed for longer life with refractory walls, more instrumentation and a process rate of about 500 pounds per hour. The results with the first pilot plant indicated that a pilot plant with a smaller capacity would yield very useful and valid data. This was an important consideration because of the material handling requirements for feed materials for larger scale pilot plants.

Initially, the second pilot plant consisted of the input system of a bin with a conveyer and a rotary air lock at the top of the reactor, the charcoal feed-out system and four flare burners for burning the off-gases. This pilot plant was modified and upgraded several times during its four years of operation. Bench scale experiments on a batch basis with a variety of lignocellulosic materials were carried out in the laboratory during this same period. The objective of these experiments was to recover the oil and gases and to obtain complete material and energy balances. The results of these tests showed that approximately 35% of the energy content of the dry feed was contained in the condensible oil present in the off-gases (Knight et al., 1974). One major modification, therefore, to the system was the addition of an air cooled condenser for recovery of the condensible oils. Other modifications included a cyclone for removal of particulates from the off-gas stream prior to the condenser, an off-gas control fan, and a refractory lined, swirl chamber for combustion of the non-condensed gases. The off-gas fan permitted the system to be operated slightly below ambient pressure. Due to the explosive nature of the off-gas stream, each component had pressure relief weighed closures as a safety feature. This second pilot plant was operated over a period of approximately four years with a variety of feed materials, which included peanut hulls, wood residues, municipal refuse, non-metallic automobile wastes, nutshells and cotton gin trash. The overall results with this wide variety of feed materials demonstrated that pyrolysis could be used in a satisfactory manner to convert the materials to useful energy forms.

A major project was conducted for Cotton Incorporated to determine the feasibility of pyrolysis of cotton gin waste as a method for disposal of the waste with maximum resource recovery. The results of this program showed that gin trash could be pyrolyzed successfully. The physical characteristics and the low bulk density (5 to 6 lbs/per cubic foot) of cotton gin trash presented material handling problems in processing gin trash through the reactor. The gin trash, which would not flow freely through the reactor, would bridge causing very erratic operation. To overcome this problem, a specially designed mechanical device was installed to agitate and to exert a downward pressure on the bed in the reactor to prevent bridge formation. The agitator solved the bridging problem, and gin trash was processed in a successful manner. Since cotton gin trash is generated seasonally, the economics of a pyrolysis system for a gin trash only are not favorable. If a pyrolysis system could be utilized the year round for a variety of wastes in a given locality, then the economics would be much more favorable.

The Environmental Protection Agency supported a pyrolysis study on the production of clean fuels from agricultural and forestry wastes. This program was a study of the operating parameters of the Georgia Tech/Tech-Air pyrolysis system, and the objective was to determine the combination of parameters that produced maximum yields of charcoal and oil with minimum yield for gas. The feed material for this study was pine sawdust. The test results showed that the maximum combined yields of charcoal and oil were obtained at the lower air-to-feed ratios. A second objective of this study was to make a prelininary design of a 200 ton per day (50% water content of the feed on a wet basis) mobile pyrolysis system for conversion of agricultural and forestry wastes into clean fuels. The results of the program indicated the technical feasibility of a mobile system.

The processing of municipal refuse was supported by the Tech-Air Corporation to determine the feasibility of the system for this type of waste. A number of pyrolysis runs were made with shredded whole municipal refuse from which most of the ferrous metals were removed and shredded light fractions of municipal refuse. The results of these tests showed that it was technically feasible to process municipal refuse with the Georgia Tech/Tech-Air system. For a detailed long term investigation of pyrolysis of municipal refuse and fractions of municipal refuse, the conclusion was made that a new pilot plant with a larger pyrolysis reactor with some different design features and more instrumentation would be required.

As a result, a third pilot plant system to process one ton/hr was designed, fabricated and assembled in the summer of 1974 for the major objective of processing municipal refuse. This system was designed with several features which were significantly different from earlier pilot plants. For the feed-in system at the top of the reactor, a double gate, input air lock was used; the interior walls of the reactor chamber were acid resistance refractory; a specifically designed stirrer was installed to level and exert a slight compaction effect on the refuse; and the output feed system was designed to pass large chunks or pieces of charcoal or inert material. This pilot plant was utilized for over a year processing various types of municipal refuse and fifty-nine test runs were made for the Tech-Air Corporation. The feed materials included whole shredded municipal refuse from which large pieces of metal had been removed; shredded light fractions of municipal refuse; shredded heavy fractions of municipal refuse; copyrolysis of the light fraction with sewage sludge; and copyrolysis of the light fraction with shredded rubber tires. These runs were successful, and the results from these tests and the processing of municipal refuse have been discussed by Bowen et al., 1978.

A study was supported by the Environmental Protection Agency to investigate three elements of a prototype mobile system for pyrolysis of agricultural wastes into clean, transportable fuels. The three elements included a study of the pyrolytic reactor itself, the operation of an internal combustion engine with simulated pyrolytic gas, and the combustion and emission characteristics of the pyrolytic charcoal and oil. From these studies and an earlier study, the dominant variable was found to be the air-to-feed ratio and that the combined energy yield of the charcoal and oil was a simple linear function of the air-to-feed ratio. The studies also indicate that feed material, converter capacity, and mechanical agitation have little influence on product yields. The results of the test of the operation of a spark-ignition internal combustion engine with dry simulated pyrolytic gas showed that excellent stability was obtained and the brake power output was 60 to 65% of that when the engine was fueled with gasoline.

A decision was made in the summer of 1975 to completely redesign and rebuild the second pilot plant. There were a number of modifications that were needed based on past operating experience and the results of a number of investigations. The redesigned system included a waste receiving bin, a belt conveyor to the feed-in-system at the top of the converter, the converter and charcoal handling system, an off-gas cyclone, a condenser, a demister, a draft fan, and a vortex burner. This rebuilt pilot plant is shown in Fig. 1. The present system will process from 300 to 1000 pounds of waste per hour depending on the density of the feed material. Types of feed materials that have been processed in this pilot plant include pinebark and sawdust mixtures, pine chips, hardwood chips, nut shells, Eucalyptus chips and Melaleuca chips.

FIGURE 1. Georgia Tech pyrolysis pilot plant, 1975 - present.

B. Development of Tech-Air Field Demonstration Units

In the overall development of the Georgia Tech/Tech-Air pyrolysis system, two parallel efforts were taking place at the same time. The efforts that were taking place at Georgia Tech have been described above which began with the initial work in a hand operated reactor and the subsequent design and construction of four pilot plant units, two of which are in current use. As mentioned above also, the process was licensed in 1970 to the Tech-Air Corporation for commercialization. The Tech-Air management made the decision in 1971 to design, construct and install two field test units, each with a nominal capacity of two tons (dry) per hour. These test units were installed at a peanut shelling plant in south Georgia. The total system had a feed storage bin and feed-in system for the peanut hulls to the converters, a charcoal feed-out system, a draft fan and stack in which the off-gases were burned. No provision was made for the condensation of any oil from the off-gas stream or for charcoal storage and handling. The two pyrolytic reactors were field tested for about one year with peanut hulls. The use of peanut hulls was a decided advantage since they had a low moisture content, and a drying operation for this feed material was not necessary. The operation of these units was carried out on a test and development basis and was successful. Significant results from this test program included an improved air injection system and the replacement of the natural gas burner start-up system with a simplified electric one. In addition, the experience gained with this system and with the pilot plant systems on campus provided valuable insight into scaleup problems and the handling of feed materials and products.

The management of Tech-Air decided in 1972 to design and build a field test system to process forestry wastes. This decision was based on the successful operation of the two ton per hour pyrolytic reactors with peanut hulls and the fact that several million tons of forestry wastes are generated yearly in Georgia as compared with about 160,000 tons of peanut hulls.

One of the needs for processing forestry wastes would be a satisfactory dryer. After an evaluation of commercial drying equipment, Tech-Air undertook the development of a compartmented, conveyor type dryer which could use waste heat or low temperature gases for drying. The field test pyrolysis facility was built and installed at a lumber mill operation in Cordele, Georgia in 1973. This facility was designed in addition to the dryer with a two ton per hour converter, an off-gas system with a cyclone for removal of particulates, a condenser for recovery of condensible oil, a draft fan, and a stack in which the noncondensed off-gases were burned. Charcoal and oil handling equipment and storage facilities were installed. The present facility is shown in Fig. 2.

The plant was operated intermittently for about a two year period with a mixture of pine bark and sawdust. Some of the accomplishments during this period were the successful development and installation of a wood waste dryer which utilized the gases from the off-gas stream for drying and successful burning of the pyrolytic oil on a demonstration basis in several commercial applications. The servicing requirements for the off-gas system were relatively high, but acceptable, for operation on a one shift basis. Prior to start up, the off-gas system could be cleaned of accumulated solids and tars for the day's operation. This servicing requirement, however, was unacceptable when the facility was operated at a later date 24 hours a day and seven days a week.

In 1974, Tech-Air with support from the American Can Company had a series of pyrolysis experiments with municipal refuse and fractions of municipal refuse. The Americology Division of the American Can Company was involved in processing municipal refuse employing a total resources recovery system. A pyrolysis system in a total resource recovery facility for municipal refuse could be used to produce more readily acceptable fuel fractions, to reduce the volume of the material, and to provide for the development of new charcoal-based products. Near the end of the testing program with municipal refuse, the American Can Company acquired the Tech-Air Corporation as a wholly owned subsidiary.

After the acquisition, the Tech-Air Corporation immediately made the decision to carry forward with the commercialization of the wood waste pyrolysis system. This required that the reliability of the 50 dry ton per day facility at Cordele for operation on 24 hour per day, seven days per week basis be established. To accomplish this, a six month program was carried out to upgrade and extend the capacity of the facility for around-the-clock operation. The dryer was completely rebuilt, a surge bin was placed between the dryer and converter, the converter was improved and the off-gas system was rebuilt. A storage facility was added for hogged feed material. An office building was added which included a laboratory for quality control. With completion of the upgrading program, plant operation was started on a 24 hour per day basis. Some of the minor problems encountered were with the material handling system which required some improvements of conveyors and the establishment of a routine serving program. The major problem, however, was in the off-gas system. As pointed out before, the off-gas system could be serviced prior to start-up on a one shift per day operation. This was not the case with

FIGURE 2. Tech-Air pyrolysis facility, 1973 - present.

24 hour per day operation as the servicing of the off-gas system resulted in considerable down time for the facility.

Based on the results of a program with the rebuilt pilot plant unit employing oil scrubbing in the off-gas system for the purpose of removing the particulates and cooling the gas stream, oil scrubbing was installed in the Cordele facility, and the on-line time was improved significantly to better than 90%. The Cordele plant was operated very successfully after the installation of oil scrubbing for several months until June, 1977. During this time, the charcoal and oil products were sold in the bulk charcoal and fuel oil markets. With this successful demonstration and operation of the facility, the reliability and value of the pyrolysis system in waste utilization and resource recovery had been established.

IV. GEORGIA TECH/TECH-AIR PYROLYSIS PROCESS

A. Description

A simplified flow diagram of the Georgia Tech/Tech-Air pyrolysis system is shown in Fig. 3. The basic elements for the process are the same for the common types of feed materials such as forestry, agricultural and municipal wastes. The particular characteristics of each feedstock must be considered from the materials handling standpoint and in developing the optimum pyrolysis conditions. The following description is for a system processing wood wastes.

The feed material is hogged so that the maximum particle size is not more than an inch or so in any dimension. Sawdust, for example, would not have to be hogged. The hogged material is conveyed to a dryer so as to dry the waste to approximately 7% moisture. The operation of the reactor is more easily controlled with feed material of reasonably constant and low moisture content. A portion of the gaseous fuel from the process is used to fire the dryer, and in the event that the amount of gaseous fuel is not sufficient, the pyrolytic oil can be used as a back-up fuel. The dried feed which is stored in a bin to provide surge capacity for the pyrolysis reactor, is fed into the top of the reactor through an airlock device and moves down through the reactor under the force of gravity. A bed-height sensing device is used to control the input into the reactor. The temperature at the top of the bed can range from approximately 350°F to 500°F. As the material moves down through the reactor, the temperature increases to the maximum in the pyrolysis zone, which can vary from 1000°F to 1700°F. The temperature in the converter

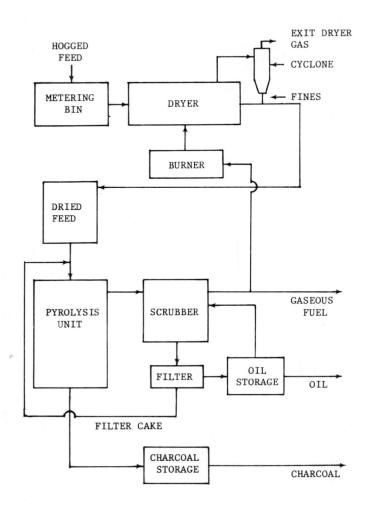

FIGURE 3. Schematic of Georgia Tech/Tech-Air pyrolysis process.

is controlled by the air-to-feed ratio. The gases produced in the reactor flow upward through the bed of descending feed material and leave the top of the bed, at temperatures ranging from 350°F to 500°F. The reactor is operated slightly under atmospheric pressure by an induced draft fan. The charcoal is discharged through the feedout mechanism at the bottom of the unit into a sealed chamber, cooled with a water spray and then conveyed to the charcoal storage bin. The rate of the charcoal discharge is used to control the throughput rate.

The off-gases from the reactor which contain noncondensible gases, condensible organics, water vapor and entrained particulates are passed into a scrubber where they are sprayed with cooled pyrolytic oil. In this scrubber-condenser, the particulates are removed and the gas stream is cooled to a temperature between 180°F and 200°F. In the condensing process, the temperature is controlled to limit the amount of water condensed with the condensible organics. The pyrolytic oil from the scrubber is filtered in a continuous filter, and the filter cake is reinjected into the feed stream for the reactor. The filtered oil is pumped into a holding tank, cooled and recirculated through the scrubber-condenser. Excess oil from the holding tank is pumped to a bulk storage tank.

The pyrolysis gases from the scrubber-condenser contain noncondensible gases, water vapor and some low boiling organics, and hence, it is desirable to utilize the gases close to the pyrolysis plant. A portion of the gases are utilized as a fuel for the dryer used to dry the feed material. The remainder of the gases can be used as a fuel for a boiler or some other heat device. A flare stack is used to burn the gases during start-up or any excess gases during operation.

The Georgia Tech/Tech-Air pyrolysis process is versatile and flexible. The distribution of the energy amoung the products—charcoal, oil and gas—can be varied by control of the operating conditions, primarily air-to-feed ratio, of the reactor, and the distribution also varies with the feedstock. The system can be operated to produce a high volatile charcoal with correspondingly reduced yields of oil and gas or a low volatile charcoal with correspondingly increased yields of oil and gas. The pyrolysis reactor can be operated as a gasifier in which the net yield of charcoal (primarily fines) is 3.8%. The off-gases from the operation of the reactor as a gasifier can be burned directly in a gas-fired boiler or the condensible portion can be condensed to obtain the pyrolytic oil which can be stored for use with peak load requirements. The system has been tested extensively with municipal refuse (Bowen et al., 1978). These tests included the successful processing of both light and heavy fractions of municipal refuse, and copyrolysis of light fraction municipal waste with both sewage sludge and shredded tires. The technical reliability of the system has been proved by the 24 hour per day several months operation of the Tech-Air Cordele plant.

The Georgia Tech/Tech-Air pyrolysis system can be adapted to meet the particular requirements for specific situations. For example, a sawmill which produces seven tons, on a dry basis, of wood wastes per hour could operate a system to produce charcoal and oil as saleable products and use the excess gas (over that required for drying the wastes) for kiln drying operations. This approach would provide for the conversion of the wood wastes into useful fuel products and eliminate the problem of disposal of the wastes. A large pulp and paper mill could operate the system as a gas and oil producer utilizing bark as feed material. The noncondensible gas would be used for base load energy requirements and the oil could be used as a fuel for peak load requirements. The gas and oil are clean burning fuels and do not require the equipment for the clean-up of stack gases as in the case with bark boilers.

B. Mass and Energy

For the mass balances with the Georgia Tech/Tech-Air pyrolysis process the feed material and air input and the output of charcoal and oil were measured directly. The quantity of the gaseous product (pyrolytic gases plus nitrogen from input air) could be determined by difference. For the energy balances, the heating values of the charcoal and oil can be determined by well known analytical techniques of representative samples of these materials. For homogeneous feed materials, such as peanut shells and wood wastes, representative samples of the input feed can be obtained for heating value determinations. With municipal refuse, which is more heterogeneous, it is more difficult to obtain representative samples of the refuse. Heating value determinations of the noncondensed gaseous product can be determined by gas chromatographic analysis of representative samples obtained with a gas-train sampling system (Knight, 1976).

The gross energy recovery in the form of the products—charcoal, oil, and gases—is approximately 95% of the energy in the dry feed as shown in Table I, which gives the mass and energy yields for pine bark and sawdust for three different cases. In case number one, the pyrolyzer was operated to obtain an increased charcoal yield. The combined charcoal and oil yields would contain approximately 68% of the input energy of the feed on a day basis. Approximately 20% of the energy

TABLE I. Mass and Energy Yields for Pine Bark and Sawdust

Product	Case 1 Charcoal yield		Case 2 Oil yield		Case 3 Gas yield	
	lb product / lb dry feed	Btu product / lb dry feed	lb product / lb dry feed	Btu product / lb dry feed	lb product / lb dry feed	Btu product / lb dry feed
Charcoal	0.35	4315	0.25	3312	0.20	2710
Oil	0.18	1634	0.23	2057	0.19	1749
Gas to process	0.20	595	0.35	1197	0.51	2100
Gas to dryer	0.58	1763	0.51	1763	0.43	1763
Total	1.31	8307	1.34	8329	1.33	8322
Percent energy recovery of input feed[a]	—	95.5	—	95.7	—	95.7

[a] Based on heating value of 8700 Btu/lb of dry pinebark and sawdust feed.

of the input feed in each case is used to fuel the dryer, which is sufficient to reduce the moisture content of wet wood with 50% moisture on a wet basis to approximately 7%. In this case, approximately 7% of the input energy is available in the gas for on-site utilization. In case number two, the charcoal yield decreased with increases in the oil and gas yields. Approximately twice as much energy is available in the "Gas-to-Process" in this case as in case one. The oil yield increased 5% over the oil yield in case one. In case three, the energy in the "Gas-to-Process" is approximately 3.5 times the gas energy yield in case one. For these cases, the average energy loss is 4.4% of the input energy. These data are very indicative of the versatility of the Georgia Tech/Tech-Air pyrolysis process and of the control that can be exercised in the operation of the system to meet the requirements for a specific situation.

The yields of the products, and consequently, the distribution of energy among the products varies with the type of feed material. Specific examples of this variation in yields are found with pine tree top chips and pine bark. With pine tree top chips, charcoal and oil yields were obtained of 20.5% and 34.6%, respectively, and these two products contained 68.6% of the input energy of the feed on a dry basis. With pine bark, the yields of charcoal and oil were 45.7% and 12.8%, respectively, and combined, would contain 79.2% of the energy of the input feed.

C. *Comparison of Georgia Tech/Tech-Air Process with Badger-Stafford Process*

The Badger-Stafford process used by the Ford Motor Company in the 1930s to process wood wastes was discussed in some detail in Section II. In view of the current interest in the utilization of wood wastes, it is instructive to compare this process and the Georgia Tech/Tech-Air process. The feed material for the Badger-Stafford retort was dried to less than 0.5% moisture and ranged in size from a match stick to pieces up to 0.75 × 2 × 8 inches. The feed material for the GT/T-A process is dried to approximately 7% moisture, and the size of the material can vary from sawdust and shredded municipal refuse to pieces not greater than approximately one inch in any dimension. The Badger-Stafford retorts used by the Ford Motor Company were 40 feet high and 10 feet outside diameter with 1.5 feet of insulation, and processed three to four tons per hour on a continuous basis. No air was introduced into the retort to sustain the reaction. Before start-up the retort was heated to 1000°F by burning gases inside the retort. The hot, dried wood feed from the driers was introduced into the top of the retort, which was operated continuously by utilizing the heat from the exothermic reaction occurring during the carbonization of the wood near the center of the retort to heat the descending wood up to the temperature necessary to initiate the exothermic reaction.

The operational mode of the GT/T-A system is different in several respects. This system requires the introduction of a small quantity of air (approximately 0.25 lb per lb feed) into the converter to sustain the reaction, and the feed material is dried to approximately 7% moisture. A converter that will process seven dry tons per hour would have an inside diameter of seven feet and be approximately 10 feet in height. The on-line time of the rated capacity of the GT/T-A system is better than 90% as demonstrated at Cordele as compared with ~67% for the Badger-Stafford units. The Ford Motor Company wood distillation facility used Seaman retorts for sawdust and shavings as relatively fine material such as sawdust could not be processed satisfactorily in the Badger-Stafford retort. The facility, in addition to producing charcoal and combustible gases, used condensers for the recovery of pitch (tar), creosote oil and pyroligneous acid from which commercial chemicals were recovered.

The operations of the Badger-Stafford retort was relatively inflexible, and hence, the product mix was fixed. This is in direct contrast to the GT/T-A process which has a high degree of versatility. A wide variety of feedstocks can be processed in the GT/T-A converter, and the operational conditions of the reactor can be adjusted to produce high volatile charcoal (high yield) with correspondingly higher yields of oil and gas; and to produce predominantly gas and oil with a char yield of 3% or less.

D. *Pyrolysis Versus Gasification*

The gasification of agricultural, forestry and municipal wastes to produce a clean burning gaseous fuel is of considerable interest and a very desirable goal. Gasifiers, a means of utilizing these wastes, are receiving a great deal of attention now since present installations using natural gas could be retrofitted to produce a gaseous fuel from waste materials. Also, gasifiers can vary in size from relatively small ones to those that could process several tons per hour. There are many opportunities throughout the country for gasifiers of wastes to produce a fuel that could be used for many different operations and processes, such as drying agricultural and forestry products and conversion to power for mechanical operations.

Pyrolysis of wastes, as has been noted, produces charcoal, oil and noncondensible gases. In gasification, however, the objective is to convert all of the waste to a gaseous fuel. This can be accomplished by increasing the air-to-feed ratio so that all of the charcoal is gasified and converted mainly to carbon monoxide. This, however, increases the temperature in the reactor because of the exothermic reaction and can lead to slagging of the ash unless proper controls are provided to prevent this from occurring. Steam can be introduced also to react with the hot charcoal to produce hydrogen and carbon monoxide and reduce the temperature. A proper balance between the exothermic and endothermic reactions that occur in the charcoal can be used to some extent to control the temperature in the charcoal portion of the reactor bed.

The use of air in the gasifier introduces nitrogen into the system which is present in the final gaseous product. The heating value of the gaseous fuel from pyrolysis systems and gasifiers which use process air falls in the range of 140 to 200 Btu per standard cubic foot. If nitrogen were not present, then the heating value would be approximately twice the above value. The hot off-gas stream from the gasifier contains water vapor and condensible organics in addition to the gaseous components. It is necessary, therefore, to utilize the gas close to the site or production and to minimize condensation of the water vapor and organics which would lead to tar formation.

The Georgia Tech/Tech-Air pyrolysis process can be operated as a gasifier in which the charcoal yield can be reduced to approximately 5% or less. The pyrolytic reactor is operated as a gasifier by screening the charcoal and reinjecting the coarse particles into the feed material. A higher air-to-feed ratio is used in normal pyrolysis operation, and the recirculation of the char allows the reactor to be operated in a nonslagging mode. In this operational mode, 90% of the input energy of the dry feed is contained in the off-gases. This off-gas stream can be used directly as a gaseous fuel if it is ducted into a boiler without allowing it to cool. Another option that is available with the operation of the GT/T-A system as a gasifier is to condense the oil from the stream and use it as a fuel oil for peak fuel requirements. In this option, the gas contains approximately 57% of the input energy of the dry feed and the oil, approximately 33%.

The Purox system, as developed by Union Carbide primarily for municipal refuse, is an example of a gasifier which uses oxygen instead of air and also is operated as a slagging gasifier (Wilson et al., 1978). The gaseous stream is subjected to cleanup and cooling to remove oil, tar, ash, particulate matter and moisture. The cleaned gas is essentially free of nitrogen and consists mainly of hydrogen, carbon monoxide, carbon dioxide and hydrocarbons, and has a heating of 370 Btu per standard cubic foot.

V. PRODUCTS AND UTILIZATION

A. Gases

The gaseous phase produced by the Georgia Tech/Tech-Air process has a low heating value on a volumetric basis, which is in the range of 200 Btu per cubic foot. The off-gases leave the top of the pyrolytic converter at a temperature ranging from 350°F to 500°F and near atmospheric pressure, and contain condensible organic oil, water vapor and noncondensible gases. The relative amounts of charcoal, oil and gas depend upon the conditions of operation. For example, as pointed out by Bowen et al. (1978), the net yield of charcoal can be lowered to 3.8% with charcoal recirculation. In this operational mode, the GT/T-A process converts the wastes to gas and oil. If all of the off-gases from the converter are ducted directly into a boiler as a gaseous fuel, then the GT/T-A process is essentially a gasifier for waste materials.

The main components of the noncondensed gaseous phase from the process are hydrogen, carbon monoxide, methane, carbon dioxide, small amounts of low molecular weight hydrocarbons, nitrogen from process air, water vapor and organic compounds. The organics consist of low boiling compounds in the vapor phase and minute liquid droplets. The composition of the off-gases will depend upon the operational parameters, and the heating value will vary, but it does not vary over a wide range. The changes in the composition and heating values of the noncondensible gases at temperatures from 540°C to 870°C from the bench scale pyrolysis of pine sawdust have been reported by Knight (1976). The heating values of the gas phase increased with increasing temperature and were in the range of 360 to 420 Btu per cubic foot. The gas phases from these experiments did not contain any nitrogen, water vapor or oil droplets. The characteristics of the pyrolysis gases are such that it is necessary to utilize these gases close to the pyrolysis plant as a fuel. The gas has a low air-to-fuel requirement for combustion as compared with the more common fuels.

B. *Pyrolysis Oils*

The oils produced from the pyrolysis of lignocellulosic materials are organic substances with a wide spectrum of organic compounds, many of which contain oxygen. The boiling range of the organic compounds vary from the low boiling volatile components to the very high boiling components. The oils are heat sensitive and begin to decompose when subjected to temperatures at approximately 200°C and above. The oils contain reactive components and the viscosity increases slowly on exposure to air. The condenser system in the GT/T-A process is operated such that the oil products have a moisture content of 10 to 15% by weight.

Two samples of pyrolysis oil from the Tech-Air Cordele facility have been extensively characterized both chemically and physically (Knight et al., 1977). The heating values of the oils are in the range of 9,100 Btu/lb, 14% water to 10,600 Btu/lb, 10.4% water. These values are lower, of course, than the heating values of petroleum fuels on a mass basis. The density of the pyrolysis oils are in the range of 9.2 to 9.5 lbs per gallon as compared with fuel oil densities of 7 to 8 lbs per gallon. On a volume basis the heating values of the oils compare more favorable with the heating values of the petroleum fuel oils than on a mass basis. The pyrolysis oils are slightly acidic, and this characteristic must be taken into account in handling and processing the oil. The oils contain essentially no sulfur and less than 1% nitrogen. The oils are soluble in organic polar solvents, such as acetone and methylene chloride, but only slightly soluble in nonpolor solvents, such as hexane. The viscosity of the oils are approximately ten times greater than number two fuel oil, but about one-tenth that of number six fuel oil. The viscosity of a sample of pyrolysis oil increased significantly when heated at 110°C for 25, 75, and 125 hours. After 75 hours, the viscosity was very similar to that of a sample of a number six fuel oil.

The most immediate and practical use for pyrolysis oils is as a fuel. The oil has been burned satisfactorily in combustion tests conducted by the Tech-Air Corporation (Bowen et al., 1978). The oil from the Cordele facility has been sold for use as a fuel in a cement kiln, a power boiler and a lime kiln. It has been fired directly and, as a 20% blend with a number six fuel oil. In the latter case, it was blended with the fuel oil just prior to firing. The water content of the pyrolysis oils lowers the viscosity and therefore improves its handling properties. It also aids in the atomization in the firing operation. Although the oil is slightly acidic, the corrosive rate is a fraction of a mil per year with 304 stainless steel. In the handling, transporting and storing of the oil, the heat sensitivity and corrosive characteristic to some metals must be taken into account. Pyrolysis oils, because of the essentially zero sulfur content, have the advantage of reducing the overall sulfur emissions when blended with sulfur-containing fuel oils. Pyrolysis oils are very satisfactory as fuel oils on the basis of available data and experience with them.

The pyrolysis oils, as stated above, contain a wide variety of organic compounds, and hence have potential as a source of materials for chemical applications. The oils contain approximately 20% phenolic compounds, which have potential for the production of resins. The Environmental Protection Agency has supported research and development programs whose objectives are to develop processes for the utilization of the oils (Knight et al., 1978) and to determine the composition of the oils (Polk, 1978). These programs will help to develop methods for processing pyrolysis oils to yield products for industrial chemical applications, and hence, enhance the economic value of the oils.

C. *Charcoal*

The GT/T-A pyrolysis process can be operated to produce a charcoal with a volatile content ranging from 3 to 25%. The measured heating values of the charcoal will range from 12,000 to 13,500 Btu per pound. For some feed materials, such as bark, the ash content of the charcoal can be high, and consequently, the heating value is less than that for a charcoal of low ash content. The bulk density of charcoal from hogged wood waste is in the range of 10 to 13 pounds per cubic foot. The charcoal from wood wastes has essentially zero sulfur and consequently does not emit any sulfur oxide emissions when burned. It is widely used in the manufacture of charcoal briquettes. The GT/T-A process is particularly well suited for producing a high volatile charcoal which is highly desirable for briquette manufacture. The charcoal can be used as solid fuel substitute for pulverized coal and, has value as a diluent for sulfur containing fuels as discussed below in Section V-D on combustion tests.

Charcoal produced from wood wastes at the Tech-Air Cordele facility and in the Georgia Tech pilot plants has been activated at the bench level and in a rotary kiln to yield a water grade activated carbon. Activated carbons are higher value products and are used in many applications which require a material with a high degree of adsorptive capacity for organic substances. The GT/T-A process is especially well suited to produce charcoals which have a high degree of activability.

D. *Combustion Tests*

A series of combustion tests were conducted by the Pittsburgh Energy Research Center (Demeter, J. J., et al., 1977) with charcoal, charcoal-coal blends, and a charcoal-pyrolytic oil-fuel oil blend. The charcoal and pyrolytic oil were produced by the GT/T-A process from wood wastes. The combustion and handling characteristics of the charcoal and charcoal-coal blends were determined in a 500 lb per hour pulverized-coal-fired combustion test facility. Combustion tests were conducted with a Pittsburgh seam coal as a base combustion test, a high volatile charcoal (28% volatiles) alone, a 50-50 blend of the high volatile charcoal and coal and a 50-50 blend of the low volatile charcoal (2.5% volatiles) and coal. The nitrogen and sulfur contents of both the high and low volatile charcoals were very low, 0.1% and the NO_x and SO_x emissions from the combustion of the high volatile charcoal were very low. The NO_x and SO_x from the blends of charcoal and coal were a weighed average of the values obtained for the fuels fired separately. The carbon combustion efficiency of 98.6% obtained with both the high and low volatile charcoal-coal blends compared favorably with the 99.1% value obtained in the coal reference test. The 97.3% obtained with the high volatile charcoal alone was attributed to the higher percentage of oversized particles.

A blend of number six fuel oil and a 60-40 mixture of low volatile charcoal and pyrolytic oil was prepared to produce a slurry containing 30% charcoal. The slurry was fired in a 100 HP oil fired, packaged firetube boiler. The flame stability in all tests was reported as excellent. The NO_x emissions were lower than those obtained from firing a coal-oil slurry, and the SO_x emissions were proportionate to the concentration of sulfur in the slurry. The results of these combustions demonstrate that the charcoal and pyrolytic oil can be fired successfully and are ideal fuels to mix with coal and petroleum fuel oils to reduce SO_x emissions and to conserve diminishing petroleum reserves.

VI. MOBILE PYROLYSIS SYSTEM

The preliminary concept of mobilizing the Georgia Tech/Tech-Air pyrolysis system was developed in the latter part of 1973 and reported by Knight et al., 1974. The need for a transportable pyrolysis system arises because many agricultural wastes are produced seasonally and other wastes are produced in limited quantities. Also, these wastes are frequently produced in widely diverse locations. It would be impractical and uneconomical to invest the necessary capital in pyrolysis facilities which would be used only for relative short periods of time during the year. Also the transportation costs to haul the waste in the form generated to a central pyrolysis facility would not be economical in many cases because of the distances involved and the bulkiness and moisture content of the wastes. The basic idea of the mobile pyrolysis system is that it could be moved from one locality to another within a given area to process wastes generated seasonally and wastes that are generated continuously but in limited quantities. In this way a system could be utilized throughout the year. Elementary calculations for wet wood waste (50% moisture on a wet basis) show that for a single truckload (net weight limited to 40,000 pounds) about 174 MM Btu of energy would be available from the wet wood waste whereas 483 MM Btu would be available from a 70% charcoal-30% oil mixture. Thus the energy transported in the form of charcoal and oil is about 2.8 times greater than that transported in the form of wet wood waste. An independent economic evaluation, which looks promising, has been performed (Wilson et al., 1978).

The Environmental Protection Agency awarded a contract in June, 1978 to Energy Resources Company, Cambridge, Massachusetts for the design, fabrication and testing of a mobile pyrolysis system. The five phases of the program are: conceptual design, detailed mechanical design, prototype fabrication, field testing and a final report with a detailed evaluation. The Energy Resources Company plans to use a fluidized bed converter of its own design. The system is to be designed to meet the EPA specifications. It is to be capable of accepting agricultural or forestry wastes in the form and moisture content at which they are generated. The system is to be easily transportable and mounted preferably on two trailers, which meet the highway regulations of California. The system is to be capable of being assembled or disassembled in one day and is to have a useful life of ten years. Compliance with all pollution control, water quality control and OSHA regulations of California and the

Federal government is a requirement for the conceptual design. The pyrolysis converter is to have a processing capacity of 100 tons of waste per day (at 10% moisture on a wet basis), and the system is to produce transportable fuel products which contain 70% of the energy content of the feedstock on a dry basis. The system, after startup, is to be capable of utilizing process gas to generate sufficient power for all power requirements of the system. The unit is to be designed to operate 24 hours per day with an uptime of 80% or better. The target date for completion of the system for field testing is June, 1979. The twelve month field testing period is scheduled to take place in California immediately after completion of the system. The initial operation and testing of the system is envisioned as a shakedown period of one to three months. The remainder of the testing period is to spend processing rice straw, lumbermill and timber residues, cotton gin wastes and cattle feedlot wastes at least at two different sites. The results of this two year program should provide the necessary data for a complete evaluation of the technical and economical feasibility of a mobile pyrolysis system.

VII. SUMMARY

Pyrolysis of agricultural and forestry residues and municipal refuse offers an excellent method for the conversion of these wastes into cleaning burning fuels. The Georgia Tech/Tech-Air pyrolysis process has been developed over the past several years. The process utilizes a vertical, gravity-fed, packed bed reactor with a countercurrent flow of gases through the reactor bed. The pyrolysis reaction is maintained in the reactor by the controlled addition of process air. The feed material is dried to less than 10% moisture on a wet basis for smooth, steady state operation. The energy recovered in the charcoal, oil and gaseous products is greater than 95% of the input energy of the feed material on a dry basis. The GT/T-A process is versatile as it can be operated to produce charcoal from high volatility to very low volatility with corresponding yields of oil and gas. The reactor can be operated to produce charcoals with a high degree of activatability. It can be operated as a gasifier by char recirculation. The technical feasibility and reliability of the process has been proved by Tech-Air by the operation of its Cordele field demonstration facility for several months 24 hours a day, seven days a week at better than 90% on-line time.

REFERENCES

Bowen, M. D., Smyly, E. D., Knight, J. A., and Purdy, K. R. (1978). Symposium on Advanced Thermal Processes for Conversion of Solid Wastes and Residues. American Chemical Society National Meeting, Anaheim, California.

Demeter, J. J., McCann, C. R., Ekmann, J. M., and Bienstock, D. (1977). Pittsburgh Energy Research Center Report PERC/RI-77/9.

Jarvis, J. P. (1959). Report on Wood Charcoal Industry in Missouri, Engineering Experiment Station, University of Missouri, Columbia, Missouri.

Knight, J. A. (1976). In "Thermal Uses and Properties of Carbohydrates and Lignins" (F. Shafizadeh, K. V. Sarkanen, and D. A. Tillman, eds.), pp. 159-173. Academic Press, New York.

Knight, J. A., Tatom, J. W., Bowen, M. D., Colcord, A. R., and Elston, L. W. (1974). American Society of Agricultural Engineers, 1974 Annual Meeting, Oklahoma State University, Stillwater, Oklahoma.

Knight, J. A., Hurst, D. R., and Elston, L. W. (1977). In "Fuels and Energy from Renewable Resources" (D. A. Tillman, K. V. Sarkanen, and L. L. Anderson, eds.), pp. 169-195. Academic Press, New York.

Knight, J. A., Elston, L. W., and Hurst, D. R. (1978). Final Report, Environmental Protection Agency, Grant No. R 804 410 010.

Nelson, W. G. (1930). *Industrial Engineering Chemistry, 22,* 312-315.

Polk, M. B. (1978). Final Report, Environmental Protection Agency, Grant No. R 804 440 010.

Simmons, F. C. (1964). *Unasylva 17,* 199-211.

Wilson, E. M., Leavens, J. M., Snyder, N. W., Brehany, J. J., and Whitman, R. F. (1978). EPA Report No. EPA-600/7-78-086, Contract No. 68-02-2101.

METHANOL FROM WOOD: A CRITICAL ASSESSMENT

R. M. Rowell

U.S. Department of Agriculture
Forest Service
Forest Products Laboratory
Madison, Wisconsin

A. E. Hokanson

Raphael Katzen Associates
Consulting Engineers
Cincinnati, Ohio

I.	INTRODUCTION	118
II.	GASIFIERS	121
	A. The Purox Gasifier	121
	B. The Moore-Canada Gasifier	123
	C. Other Gasifier Designs	124
III.	GAS PURIFICATION	125
IV.	SHIFT AND SYNTHESIS REACTORS	129
V.	ECONOMICS	130
	A. Plant Size	131
	B. Capital Investment	131
	C. Operating Costs	131
	D. Comparison with Natural Gas and Coal	134
VI.	WOOD RESIDUE INVENTORY	136
VII.	CONCLUSIONS	142

I. INTRODUCTION

To reduce petroleum consumption in the United States, one proposal is to blend 10 to 15% of methyl alcohol (methanol) in gasoline for automotive use (Reed and Lerner, 1973; *Chemical Week,* 1977). At a current gasoline consumption rate of approximately 110 billion gallons per year (gpy), 11 to 16.5 billion gallons (gal) of methanol would be required. The 1978 methanol capacity in the United States totals 1.42 billion gpy (*Chem. & Eng. News,* 1978) of which 99% is petroleum-derived, either from natural gas or from refinery light-gas streams. To increase methanol production, recommendations have been made to use other sources of carbon for synthesis of methanol. Coal, lignite, peat, biomass, and municipal solid waste have all been suggested as possible carbon sources.

Several technical and economic feasibility studies for producing methanol from nonpetroleum source have been published. Using North Dakota lignite as a carbon source, Wentworth (1977) gives the cost of methanol at about 20 cents per gallon (cpg) for a 25,000 tons per day (tpd) fuel-grade methanol plant. Burke (1975) gives data on a 5,000 tpd methanol plant using New Mexico coal. With a plant cost of $573 million (1975 dollars), methanol price would be 41 cpg for $5 per ton coal and 63 cpg for $9 per ton coal. A report from the National Academy of Sciences (1976) estimates that 11,690 tpd of coal could produce 5,000 tpd methanol. With a capital investment of $240 million and $20 per ton coal, methanol cost would be 25.4 cpg.

A study by the city of Seattle (1974), using municipal solid waste as a carbon source, a 300 tpd methanol plant costing $56 million (1974 dollars) could produce methanol at about 43 cpg. This was estimated on a no-cost raw material basis. Barr and Parker (1976) estimate that 140 million tons of municipal solid waste are available every year in the United States; at a conversion efficiency of 19%, this source could yield 26 million tons per year (tpy) or 7,900 million gpy. They also estimate that 28 million tpy of feed lot manure could be converted at an efficiency of 7.5% to yield 2.1 million tpy or 620 million gpy.

With 5 million acres of forest infested with spruce budworm, the state of Maine (1975) speculated that this resource of approximately 208 million cords or 312 million tons could be converted with an efficiency of 28% to 89 million tons or 27 billion gal methanol. Two Canadian studies (*Canadian Chemical Processing,* 1977) report the feasibility of converting wood to methanol. In one a 990 tpd methanol plant would require a capital investment of $170 million; with a conversion efficiency of 32%, the resulting methanol would cost $.50 to $1.04 per gal. In the second study, a 305 tpd wood processing plant with a capital cost of $65.3 million (1975 dollars) and a raw material cost of $18 per ton, methanol would cost about 40 cpg.

The National Academy of Science (1976) report describes a 900 tpd wood waste processing plant producing 300 tpd methanol would have a capital cost of $29.5 million. The cost of methanol ranged from 20 cpg for free raw material to 52 cpg for wood at $35 per ton. Barr and Parker (1976) estimate that the total unutilized mill wastes, sawdust, and logging residues (130 million tons) from the United States could be converted to 38 million tons or 11,500 million gal methanol per year.

The Forest Service, in cooperation with the National Science Foundation and the Federal Energy Administration, conducted a technical and economic feasibility study for producing methanol from wood waste (Forest Service, 1976; Hokanson and Rowell, 1977).

The concept of producing methanol from wood is not a new one. Methanol obtained from the destructive distillation of wood was the only commercial source until the introduction of the synthetic process in 1927. The production of methanol since 1930 is given in Table I. Yields of about 1 to 2% or 6 gallons per ton (gpt) were achieved from hardwoods. The yield from softwoods was about half that of hardwoods. In the 1930s, individual methanol plants ranged in size from 20 to 40 tpd. By the early 1950s, the average plant size had grown to 1500 to 2000 tpd. It has been proposed that a single methanol train of 5000 tpd will be on line in 1979.

About 50% of the methanol produced goes into the production of formaldehyde; 10% into solvents; 10% into acrylics; 10% into insecticides; 10% into fungicides; 5% into textile fibers; 2% into miscellaneous uses, and about 3% is exported (Chemical Economics Handbook, 1971). The large use of methanol for formaldehyde coupled with the large use of formaldehyde in plywood adhesives links new housing starts directly to the production of methanol.

Methanol is produced synthetically from carbon monoxide (CO) and hydrogen (H_2). In a catalyst-filled converter operating at pressures ranging from 1500 to 4000 pounds per square inch (psi), two volumes of hydrogen to one volume of carbon monoxide react to form a crude methanol which is refined. In today's methanol plants, natural gas, consisting primarily of methane, is steam-reformed catalytically into CO and H_2. A small amount of carbon dioxide is added to the

TABLE I. Production of Methanol (1000 tons)[a]

Year	Natural	Synthetic	Total	Capacity
1930	12.9	25.1	38.0	–
1935	13.4	59.7	73.1	–
1940	14.1	148.8	162.9	–
1945	9.4	246.1	255.5	–
1950	6.9	450.0	456.9	–
1955	7.3	670.6	677.9	–
1960	7.2	981.2	988.4	1093
1965	–	1433		1440
1970	–	2468		2879
1975	–	2584		–
1976	–	3121		4305
1977	–	3229		4702

[a] Hagen 1976, Chemical and Engineering News, January 9, June 12, 1978, to convert to: U.S. gal multiplied by 302 gal/ton; metric tonnes divide by 1.1 U.S. ton/tonne.

methane to permit part of the hydrogen to form additional CO so that the final gas product contains two volumes of H_2 to one volume of CO. These reactions are:

$$3CH_4 + CO_2 + 2H_2O \rightarrow 8H_2 + 4\ CO$$

$$8H_2 + 4\ CO \rightarrow 4\ CH_3OH$$

Any carbonaceous material such as coal, lignite, or wood waste can be utilized for synthetic methanol production. However, in contrast to natural gas, these rat materials require several additional processing steps to refine the crude gas product into a final, clean gas product (syngas) consisting of two parts of H_2 to one part of CO. As a result, the conversion of a carbonaceous material is considerably more energy-intensive than that required by natural gas.

Its logistics are considerably greater (solids handling vs. pipeline). Its yield is less. A schematic drawing of the overall process steps for converting wood waste into methanol is shown in Figure 1.

II. GASIFIERS

For any solid carbonaceous material to be converted into a syngas, it is first necessary to produce a crude gas consisting primarily of H_2, CO, and CO_2. Gasification by partial oxidation is the common design system. If air is used to oxidize the feed material, the crude gas contains about 46% nitrogen, which can be removed by cryogenic means. If oxygen is used instead of air, a cryogenic system is required for initial separation of air into oxygen and nitrogen. Also, about 2% of the wood (dry basis) is converted to an oil-tar fraction. Several types of gasifiers have been developed for the partial oxidation of wood, wood waste, and garbage. These are designed to operate at atmospheric pressure, in contrast to coal gasifiers which can operate at pressures up to 400 pounds per square inch gage (psig). A comparison of the crude gas from two types of gasifiers is shown in Table II.

A. The Purox Gasifier

Under the name "Purox," Union Carbide has developed a process for the partial oxidation of preferably shredded municipal solid waste using oxygen (Anderson, 1973). The reactor-gasifier employs a moving bed in which pure oxygen is used to partially combust char into CO and CO_2, which then rise through the material being reacted. As the material flows through the reactor, it passes successively through stages of drying, reduction, and char oxidation; at the bottom, ash is removed in molten form at a temperature of about 3000°F. Crude gas containing a large amount of moisture leaves the top at a temperature of about 200°F.

Development work was initiated by Union Carbide in 1970 in a pilot plant in Tarrytown, New York sized for a feed rate of 5 tpd of municipal solid waste. In 1974, a demonstration plant in South Charleston, West Virginia, using a 10.25-foot diameter reactor, having a capacity of 200 tpd of municipal solid waste, was put into operation. Scaleup of the reactor to this level resulted in higher levels of hydrocarbons and carbon dioxide production.

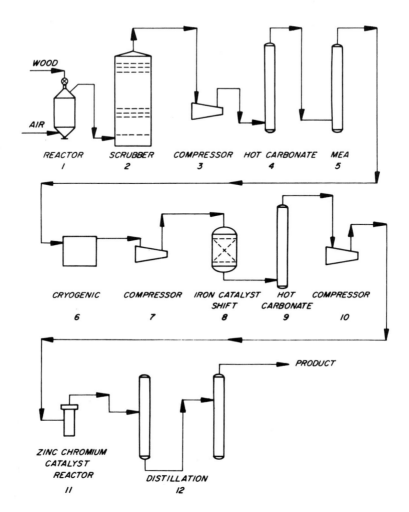

FIGURE 1. Methanol synthesis from wood waste. Process steps: (1) Partial oxidation of wood waste, (2) clean and cool crude gas, (3) compress to 100 psig, (4) remove carbon dioxide, (5) remove residual carbon dioxide, (6) remove nitrogen and hydrocarbons, (7) compress to 400 psig, (8) shift gas to two parts hydrogen and one part carbon monoxide, (9) remove carbon dioxide formed in shift, (10) compress to 2500 psig, (11) convert hydrogen and carbon monoxide into methanol, and (12) refine crude methanol into specification grade product.

TABLE II. Comparative Crude Gas Composition (dry basis)

Raw gas component	UCC Purox % (Vol.)	Moore-Canada % (Vol.)
Hydrogen	26.0	18.0
Carbon monoxide	40.0	22.8
Carbon dioxide	23.0	9.2
Methane	5.0	2.5
Hydrocarbon	5.0	0.9
Oxygen	0.5	0.5
Nitrogen	0.5	45.8
Total	100.0	100.0

Although the feed material for the UCC Purox reactor is municipal solid waste, it is expected that the crude gas composition will be essentially the same for wood waste because municipal solid waste has been found to have a similar composition with regard to carbon, hydrogen, and oxygen.

B. The Moore-Canada Gasifier

Moore-Canada of Richmond, British Columbia has developed a moving bed reactor for producing a low Btu gas from "as-is" wood waste. In contrast to Purox, the Moore reactor relies on the use of air as the oxidizing medium. Because of the high nitrogen content, the raw gas has a heating value of about 180 British thermal units per standard cubic foot (Btu/scf) in contrast to that of the heating value of the Purox unit of 350 Btu/scf.

Currently, a semi-works unit is in operation. This unit has a 5.5 foot-diameter gasifier with a capacity of about 18 ovendry tons per day (otpd) of wood waste. A commercial installation was put into operation in 1976. This facility utilizes two 9.5 foot-diameter reactors, each capable of processing 60 otpd of hogged wood waste.

Operation of the Moore reactor is similar to Purox in that the feed material enters at the top and the wood ash is discharged from the bottom. However, because air rather than oxygen is used, the maximum temperature of the oxidation (lower) zone is only about 2200°F; waste is discharged as a solid in a granular form rather than as a molten slag. Pressure at the base of the reactor is approximately 6-8 psig and, at the top, 2-3 psig. The raw gas leaves the reactor at a temperature of 160° or 180°F. By adding steam to the air, the hydrogen content of the crude gas from the Moore reactor-gasifier is increased from 8 to 10% to 18 to 22%.

C. Other Gasifier Designs

Battelle—The Battelle Pacific Northwest Laboratories carried out a 1-year pilot plant study on the partial oxidation of municipal refuse in a 3-foot (inside diameter) moving bed reactor pilot plant (Hammond, 1972). This study also included partial oxidation of wood chips using air and steam with results that approximated those reported by Moore-Canada.

Thermex—A 50 otpd demonstration plant for the gasification of wood waste has been put into operation in Edmonton, Alberta, Canada, by Alberta Industrial Development, Ltd. Under its present mode of operation, it produces a char and a low-Btu gas, but it is understood that it can be designed to operate without forming char. The gasifier is a fluidized-bed type requiring that the wood waste feed be hammermilled to less than 2-inch particle size. No data are available at this time regarding gas composition.

Copeland—The Copeland organization has built, over the past 15 years, a number of fluidized-bed reactors for the pulp industry for disposal of the organic matter in waste liquor. Such a unit would be able to accept "as-is" wood waste and sludge, but its applicability to partial oxidation for syngas has not been investigated.

Lurgi—The Lurgi reactor is designed to gasify coal with either air or oxygen and steam at 300-400 psig (Maddox, 1975). Gas values vary from 150 to 300 Btu/ft^3 depending on oxygen source. It can handle only noncaking type coal, utilizing a particle size range from 3/8 to 2 inches. At this time, no attempt has been made to process wood waste in a Lurgi reactor. The reactor requires relatively uniform particle size and would not be expected to handle wood waste.

Winkler—The Winkler unit is a fluidized bed coal gasifier operating on pure oxygen at or near atmospheric pressure (Maddox, 1975). These gasifiers are typically 18 feet in diameter and operate at a temperature of about 2200°F. Coal fed to the Winkler is ground to less than 1/4 inch. No attempt has been made to apply the Winkler gasifier to wood waste, nor does it appear to offer promise in this area since particle size is limiting.

Koppers-Totzek—These entrained flow gasification units process pulverized (-200 mesh) coal with steam and oxygen under slagging conditions at atmospheric pressure at temperatures up to 3500°F (Maddox, 1975). Although 17 Koppers-Totzek installations have been built, employing a total of about 40 gasifiers, this type gasifier is not deemed practical for the handling of wood waste because of requirements for finely ground feed.

A summary of five of these gasifiers is shown in Table III. On the basis of currently available data, the Moore-Canada gasification system comes closest to being an on-line commercial technology directly related to wood processing. For this reason, the Moore-Canada system was adapted as the basis for this study.

The material balance for the Moore-Canada gasification system is shown in Tables IV and V. Twenty-six reactor gasifiers of the Moore-Canada type are required to process 1500 otpd of wood waste. In order to provide 10 days' wood storage, 3 million cubic feet (cf) of wood waste must be accumulated. The waste is fed into the gasifiers from hoppers with a rotary gate in the feed chute to minimize back-leakage of gas. Steam and air are admitted at the base of the reactor. In addition, the decanted tar from the raw gas scrubbing system is fed back into the reactor for oxidation into additional gas.

III. GAS PURIFICATION

Crude gas from partial oxidation units is processed to remove water vapor, tars, organics, hydrocarbons, nitrogen and CO_2. The clean gas, containing primarily H_2 and CO, is then processed in a shift reactor to react part of the CO to form additional H_2, so the final gas contains the proper ratio of 2:1 of H_2 and CO. in the shift reactor, additional CO_2 is formed, thus making it necessary to again scrub the gas before the synthesis reactor.

Crude gas from the gasifiers passes upward in the single cooler-absorber-scrubber, cooling the gas from about 180° to 90°F in three stages of contacting (Fig. 1, step 2). In the lower two stages, cooled, recirculated liquor streams contact the crude gas. In the upper stage, reclaimed water condensate is used to complete the removal of organic compounds such as acetic acid.

TABLE III. Types of Reactor-Gasifiers

Item	Moore semi-works	Moore full-scale	UCC purox	Battelle pilot	Thermex semi-works
Type reactor (bed)	Moving	Moving	Moving	Moving	Fluidized
Feed	Wood waste	Wood waste	Municipal waste	Wood	Wood waste
Form of feed	Hogged	Hogged	Hogged	Chips	Hammer-milled
Air/oxygen	Air	Air	Oxygen	Air	Air
Feed rate, lb/h ovendry	1500	5000	14,000	1600	4000
Diameter, ft	5.5	9.5	10	3	8
Mass velocity, lb/h ft^2	63	70.5	123	140	80
Temp. at bottom, °F	2200	2200	3000	–	–
Form ash	Granular	Granular	Molten	–	–

TABLE IV. Material Balance for Moore-Canada Feed to Reactor-Gasifier

Unit	Pounds per ovendry ton of wood	Pounds per hour
Wood waste		
62.5 ot/h		
125,000 lb/h		
1,500 otpd		
Carbon	1,016	63,500
Hydrogen	126	7,880
Oxygen	796	49,740
Nitrogen	2	130
Ash	60	3,750
Moisture	2,000	125,000
	4,000	250,000
Air		
Oxygen	750	46,875
Nitrogen	2,475	154,570
Water	55	3,455
	3,280	204,900
Steam		
Water	333	20,800
Total in		475,700

TABLE V. Material Balance for Moore-Canada Products from Reactor-Gasifier

Unit	Molecular weight	Moles per hour	Pounds per hour
Crude gas[a]			
Hydrogen	2	2,662	5,324
Carbon monoxide	28	3,325	98,100
Carbon dioxide	44	1,237	54,430
Methane	16	363	5,810
Hydrocarbon—average	41	131	5,370
Nitrogen	28	5,525	154,700
Oxygen	32	69	2,206
Moisture	18	438	7,885
		13,750	328,825
Ash			3,750
Condensate—water layer			
Organic compounds			3,600
Water			139,525
			143,125
Total out			475,700

[a] Following partial condensation and separation of condensate.

Because the moisture condensed from the crude gas (roughly equal in weight to the dry wood substance entering the system) contains about 2% of soluble organics, it is necessary to clean the steam for environmental reasons. One technique for organics recovery is with a suitable solvent such as methyl ethyl ketone in a liquid-liquid multiple-stage extraction operation. The extract or light density phase is processed in an extraction tower to recover the solvent overhead and the organic-rich material from the bottom. The heavy density raffinate phase is processed in a raffinate stripper to recover that portion of the solvent dissolving in the water phase. From the bottom of the raffinate stripper, the effluent is essentially a water product of low biochemical oxygen demand. The organic product from the extraction tower may be used as a fuel in the boiler. Alternatively, it may be economically feasible to separate the stream into its components, mostly acetic acid, for byproduct value.

The cooled and partially purified gas is then compressed to about 100 psig (Fig. 1, step 3) and treated in a two-stage system to remove carbon dioxide. In the first stage, a hot potassium carbonate system is used to reduce CO_2 content to about 300 parts per million (ppm) (Fig. 1, step 4). In the second stage, it is reduced to about 50 ppm using monoethanolamine as a scrubbing agent (Fig. 1, step 5). A single-stage system of monoethanolamine could be used, but at the expense of greatly increased steam consumption.

The clean compressed gas passes to a cryogenic system (Fig. 1, step 6). In a series of switching exchangers, the residual CO_2 and water vapor are removed to prevent freezeup in the downstream exchangers and distillation towers. Next, methane and hydrocarbons are removed. Cryogenic distillation is used to separate CO from nitrogen; the liquid nitrogen leaving the system is used to precool the incoming gas. The purified product gas is a mixture of carbon monoxide and hydrogen. However, it requires further processing because it is not in the ratio of 2:1 of H_2 and CO required for syngas to produce methanol.

IV. SHIFT AND SYNTHESIS REACTIONS

Following cryogenic separation of the "inerts," the gas is compressed to 400 psig for shift conversion. A portion of the CO reacts with water vapor in the presence of an iron catalyst to form additional hydrogen (Fig. 1, step 8), to the extent that the final gas contains the required 2 parts of hydrogen to 1 part of carbon monoxide. The following exothermic reaction takes place in the shift reactor:

$$CO + H_2O \xrightarrow{catal.} H_2 + CO_2 \qquad \Delta h = -590 \text{ Btu/lb } CO$$

Because the shift reaction produces carbon dioxide, it is necessary to utilize the hot potassium carbonate absorption system (Fig. 1, step 9) which removed about 97% of the carbon dioxide formed during the shift reaction.

The synthesis gas is now compressed to a pressure ranging from 1500 to 4000 psig and fed into the methanol synthesis reactor. In the reactor, approximately 95% of the gas is converted to methanol, the balance passing as inerts to the boiler. This exothermic reaction is as follows:

$$2H_2 + CO \xrightarrow{catal.} CH_3OH \qquad \Delta H = -1200 \text{ Btu/lb } CO$$

Two processes are available for synthesis of methanol, the Vulcan process using a zinc-chrome catalyst operating at pressures ranging from about 2000 to 4000 psig, and the more recent ICI copper catalyst process operating at pressures ranging from 1000 to 2000 psig. A key factor in setting the system pressure for either process is the purity of the feed. With increasing amounts of impurities, the system requires higher pressure to minimize the effects of "inerts."

The crude methanol product from the synthesis reactor passes to a distillation train for separation of the light ends and higher alcohols from the methanol product. The mixture of light ends and higher alcohols is used as a fuel in the boiler. Catalyst life is expected to be 6 years for methanol synthesis and 2 to 3 years for the shift reactor.

V. ECONOMICS

In a chemical processing facility, production cost depends mainly on capital investment and raw material cost. The most common means for reducing unit costs is to build a high-capacity facility because investment for scaleup generally increases as production increases by a 0.6 factor. In recent years, the trend in the synthetic methanol industry has been to increase plant sizes from about 50 million gpy to 200 million gpy.

A. Plant Size

A wood-waste plant with a capacity of 50 million gpy of methanol, comparable to a small size plant making methanol from natural gas, was selected for this study. The investment for this size facility can be scaled up in accordance with standard procedures used by engineering organizations for chemical processing plants.

B. Capital Investment

An estimate of investment and operating costs for a 50 million gpy synthetic methanol plant from wood waste has been developed. To determine the effect of scaleup on cost, an estimate has also been prepared for a facility capable of producing 200 million gpy. These estimates are compared with estimates of investment and operating costs for facilities using natural gas and coal as raw materials. All of these investments are based on facilities utilizing boilers to produce steam to generate electricity and to drive turbines required for compression. Therefore, these plants are self-sufficient, requiring no outside utilities other than cooling water makeup.

The investment estimate requirement for a 50 million gpy methanol plant using wood waste totals $64 million (1975 dollars). A breakdown of this estimate into key sections is given in Table VI.

This estimate covers a complete "grass roots" facility, including offsite utilities, wood yard handling facilities, finished product storage, and office and laboratory buildings. It includes a contingency of 25% and working capital of 5%. No provision has been made for expected escalation in cost of equipment and construction labor.

C. Operating Costs

Raw material costs are the most significant operating costs. The Forest Service has estimated that the cost of collecting and transporting wood waste to a central location for processing would range between $15 and $34/ot. For comparison, the current delivered price for pulpwood ranges between $30 and $60/ot. Aside from its value as a potential source of methanol, wood waste may be used as a source of fuel. Its minimum value would thus seem to be its fuel value relative to more conventional fuels. In the 1975 fuel market, wood waste can compete with other fuels when its

TABLE VI. Investment Estimate for Wood Waste Methanol Plant of 50 Million gpy

Description		Investment
Wood yard		$3,010,000
Gasifier		4,000,000
Cooling/scrubbing		802,000
Organic recovery		2,256,000
Primary compression		1,597,000
CO_2 removal		1,704,000
Cryogenic		4,000,000
MeOH synthesis		7,330,000
Offsites		
Steam generation	$7,200,000	
Electrical generation	1,587,000	
Water treating	840,000	
Cooling tower	1,727,000	
Storage and shipping	1,152,000	
Fire protection	250,000	
		12,756,000
Distribution		
Steam	$ 832,000	
Water	1,360,000	
Electric	875,000	
		3,067,000

(Continued)

TABLE VI. *(Continued)*

Description		Investment
Buildings/structure		
Administration	$ 362,000	
Laboratory	106,000	
Maintenance	348,000	
Control house	104,000	
Compressor building	105,000	
Utility building	300,000	
		$1,370,000
Site development		
Clearing, grading	$ 75,000	
Roadway, parking	276,000	
Fencing	36,000	
Railroad siding	80,000	
Sewer facilities	40,000	
		507,000
Engineering and license fee		6,360,000
Estimate subtotal		$48,759,000
Contingency—25%		12,191,000
		$60,950,000
Working capital—5%		3,050,000
Total investment		$64,000,000

price is below $24/ot, so that this would set the minimum cost for waste wood utilized in a chemical process. A comparison of wood waste with conventional fuels, assuming a value of $2/million Btu of net heat recovered, is shown in Table VII.

An estimate of operating costs for the 50 million gpy methanol plant is listed in Table VIII. Production costs include fixed costs, raw material, labor, and overhead. Fixed costs are based on an allowance of 8% for depreciation, 4% for maintenance (including labor and material) and 2% for local taxes and insurance. Profit is based on a nominal return of 30% on investment, equal to 15% after Federal income tax.

At a wood waste cost of $34/ot, the selling price of methanol is estimated at $0.96/gal. At a wood waste cost of $15/ot, the delivered price is $0.77/gal. The 1975 price of methanol was $0.38/gal, Gulf Coast, tank car lots. The 1978 price of methanol is $0.44.

D. *Comparison with Natural Gas and Coal*

Because of the simplicity of the conversion of natural gas to methanol, the investment cost for such a plant is about one-third that of a comparable wood waste facility. Also, conversion efficiency of natural gas to methanol is significantly greater than that of wood waste. It takes

TABLE VII. Comparison of Fuel Values of Wood Waste, Coal, Oil, and Natural Gas

Source	Gross heating value	Combustion efficiency	Fuel value
Wood waste	18 million Btu/ot	66%[a]	$24.00/ot
Coal	24 million Btu/ot	80%	$38.40/ot
Oil	6 million Btu/bbl	85%	$10.20/bbl
Natural gas	1 million Btu/mcf	87%	$ 1.75/mcf

[a] Based on "as-is" or about 50% moisture by weight.

TABLE VIII. Production Cost Estimate and Product Price, Wood Waste Methanol Plant of 50 Million Gpy

	Annual (Million dollars)	Dollars per gallon	Percent
Fixed costs			
Depreciation 8% investment	5.12		
Maintenance 4% investment	2.56		
Taxes and insurance 2% investment	1.28		
	8.96	0.179	18.7
Raw material			
Wood waste—1,500 tpd at $34/ot	17.36	0.347	36.2
Labor			
Operators—10 stations at $80,000/yr	0.80		
Foremen—3 stations at $100,000/yr	0.30		
Management— at $100,000/yr	0.10		
	1.20	0.024	2.5
Overhead—100% labor	1.20	0.024	2.5
Profit—30% of investment before taxes	19.20	0.384	40.1
Total	47.92	0.958	100.0

150 cf of natural gas (containing more than 95% methane) or 4.9 lb to make 1 gal methanol. A comparison of efficiency of conversion of natural gas, coal, and wood waste is shown in Figure 2.

Conversion of coal to methanol, while considerably more efficient than that of waste wood, involves more processing facilities because of the greater amount of ash and sulfur (wood has no sulfur). Coal conversion to syngas is more efficient because it has a higher carbon content and less oxygen than wood. Comparison of investment requirements for a 50 and 200 million gpy methanol facility for each of these raw materials is shown in Figure 3.

The raw material input for three types of methanol synthesis plants is shown in Table IX. It is of interest to note that the 50 million gpy facility utilizing 1500 otpd wood waste is comparable to a pulpmill producing about 800 tpd of finished pulp. The 200 million gpy facility would be comparable to 3300 tpd production of pulp. Methanol selling price was calculated on a basis of 30% annual profit on investment, 15% after Federal income tax. Production costs, gross profit, and net profit for each type of facility is shown in Table X. A comparison of the delivered price of methanol as a function of the raw material costs and plant size for these three methanol facilities is shown in Figure 4.

VI. WOOD RESIDUE INVENTORY

The yield of methanol from wood by the gasification process is about 100 gal odt. This yield includes the production energy required to make the methanol. Forest Service data (Wahlgren and Ellis, 1978) has estimated the total potential wood residue available in the United States. Commercial forest lands contain about 20 to 30 billion tons of aboveground wood, bark, and foliage. This cellulosic material is roughly made up of 80% wood, 12% bark, and 8% foliage.

Table XI gives an estimate of U.S. wood resources. The figures for cull and salvable dead represent rough and rotten trees, residues from harvesting operations, trees from timber stand improvements, land clearing, and changes in land use.

It is difficult to estimate a realistic amount of wood residue that could be used to produce methanol. Some of it is low volume per acre material, some is located long distance from potential use sites, and some is located in inaccessible areas where the removal costs would be prohibited. The material available at a cost of $10 to $30 per ton probably represents less than half the estimated tonnage (Wahlgren and Ellis 1978).

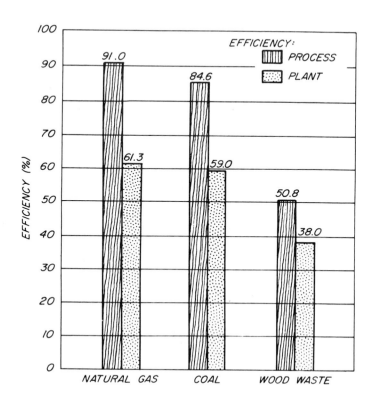

FIGURE 2. Methanol synthesis conversion efficiency for natural gas, coal, and wood waste.

Heating value: Natural gas—1000 Btu/cf; coal, New Mexico coal with 19% ash content—8600 Btu/lb; wood waste, Douglas-fir with 25% bark content—9000 Btu/lb.

Efficiency: Process efficiency—heating value of methanol as a percent of heating value of process feed.

Plant efficiency: Heat value of methanol as a percent of total energy input into plant.

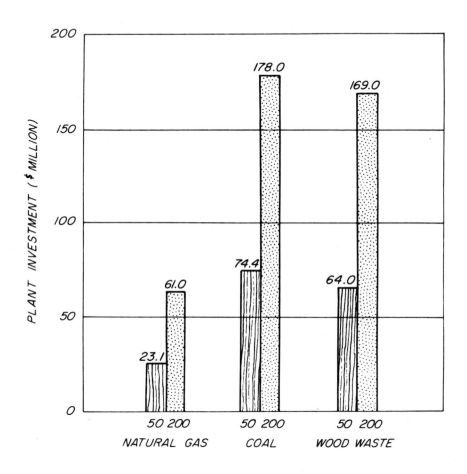

FIGURE 3. Methanol synthesis plant investment for plants of 50 and 200 million gpy facilities (additional cost for coal vs. wood due to pressurized system, increased steam requirements, and desulfurization equipment). Investment estimate: Based on 1975 costs; includes 25% contingency; no escalation included.

TABLE IX. Raw Material Input for Methanol Plants

Source	50 Million gpy	200 Million gpy
Natural gas	16.3 million cfd	65.2 million cfd
Coal	1380 otpd	5520 otpd
Wood waste	1500 otpd	6000 otpd

TABLE X. Methanol Selling Price

	Cents per gallon			
	Production cost	Gross profit	Net profit	Selling price
50 million gpy				
Natural gas at $1.75/mcf	32.0	14.0	7.0	46.0
Coal at $38/ton	53.4	44.6	22.3	98.0
Wood waste at $34/ot	59.6	38.4	19.2	98.0
200 million gpy				
Natural gas at $1.75/mcf	25.8	9.2	4.6	35.0
Coal at $38/ton	41.4	26.6	13.3	78.0
Wood waste at $34/ot	57.8	25.2	12.6	83.0

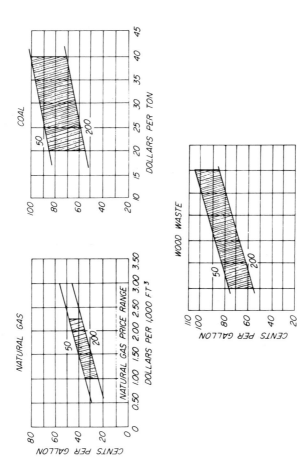

FIGURE 4. Methanol 1975 selling prices from natural gas, coal, and wood waste with plant capacities of 50 and 200 million gpy. Production costs/Percent investment: Depreciation, 8%; maintenance, 4%; taxes and insurance, 2%; profit before taxes, 30%; profit after taxes, 15%.

TABLE XI. Distribution of United States Forest Resources[a]
(Billions of tons, ovendry basis)

	Total	Sawtimber	Poletimber	Cull and salvable dead	Saplings
Wood					
Main stem	15.4	7.4	3.2	1.4	3.4
Branches and tops	4.0	1.7	0.9	0.4	1.0
Bark	3.5	1.4	0.8	0.3	1.0
Foliage	2.1	0.7	0.5	0.1	0.8
Total above ground	25.0	11.2	5.4	2.2	6.2

[a] Based on data from: USDA Forest Service (1973). "The Outlook for Timber in the United States," Forest Resource Report No. 20, 367 pp.
Keays, J. L. (1975). Biomass of Forest Residuals; AIChE; 71(146): 10-25.
Data may not add to totals because of truncating.

VII. CONCLUSIONS

It is technically feasible but not economically attractive now to produce methanol from wood residues. A methanol-from-wood waste facility having a capacity of 50 million gpy requires 1500 otpd. The yield of methanol from wood is about 38% or about 100 gal/ot. This yield is based on all process energy required coming from the wood residues. At a wood residue cost of $15/ot, the selling price of methanol is estimated at $0.77/gal; at $34/ot, the selling price is $0.96/ot. The current selling price of methanol is $0.44/gal.

If the use of natural gas is prohibited for the production of methanol, a possible source for syngas would be refinery light-gas streams. There has been a great deal of discussion recently about using the vast amount of Middle East flair gases to synthesize methanol. Another large potential source for methanol comes from the vast coal supply in the U.S., particularly for a methanol plant located near or adjacent to a mining operation. For this purpose, high-sulfur coal could be used. Finally, the deposits of lignite offer another large potential source of methanol.

No methanol has yet been produced from the gasification of wood. For this reason, any study on methanol-from-wood is based on technology that was developed for another resource. In carrying out this Forest Service technical study, known on-line technology was used. The Moore-Canada gasifier system was used because, at the time, it was the only gasifier designed to burn wood. The Union Carbide Purox unit looks very promising, but was designed to gasify municipal waste. A key to improving the efficiency of the total process of methanol from wood lies in improving the efficiency of the reactor-gasifiers to increase CO and H_2 contents and thereby reduce CO_2 and hydrocarbons. Particle size of gasifier feed materials and water content need to be optimized for greatest reactivity. Pressure partial oxidation should be investigated to determine its practicability and efficiency. Rapid changes in gasifier technology are taking place today and future development work will increase overall process efficiency.

The 1500 otpd wood residues unit used in this study is about the same quantity of wood required for a kraft pulpmill producing 800 tpd of pulp. In view of the pulp industry's progress in utilizing lower and lower grade feed materials such as sawdust and some bark, wood residues might be utilized in the near future to produce pulp. Pulp would have a product value about three times that of wood residues converted to methanol.

The energy yields for methanol from wood residues are relatively poor. Even in the case of natural gas, the final methanol product contains about 60% of the total energy input into the plant. For a wood residue facility, the yield is about 38%. That is, the total energy input for a 50 million gpy methanol unit utilizing 1500 otpd amounts to 1125 million Btu/hr and the final methanol product has a heat value of 427 million Btu/hr (gross basis).

REFERENCES

Anderson, J. E. (1973). "Solid Refuse Disposal Process and Apparatus." U.S. Patent No. 3,729,298, April 24.

Barr, W. J., and Parker, F. A. (1976). "The Introduction of Methanol as a New Fuel into the United States Economy." Foundation on Ocean Research, March.

Burke, D. P. (1975). *Chemical Week,* September 24, 33-42.

Canadian Chemical Processing (1977) *61*(6), 40-42.

Canadian Chemical Processing (1977) *61*(8), 76-79.

Chemical and Engineering News (1978). January 9, 13.

Chemical Economics Handbook (1971). Stanford Research Institute, Section 674.5022 T.

Chemical Week (1977). November 30, 28.

City of Seattle (1974). "Feasibility Study—Conversion of Solid Waste to Methanol or Ammonia," September 6.

Forest Service, USDA (1976). "The Feasibility of Utilizing Forest Residues for Energy and Chemicals." National Technical Information Service Publication PB 258-630.

Forest Service, USDA (1976). "Chemicals from Wood Waste." National Technical Information Service Publication PB 262-489.

Hagen, D. L. (1976). "Methanol: Its Synthesis, Use as a Fuel, Economics, and Hazards." Master of Science Thesis, University of Minnesota.

Hammond, V. L. (1972). "Pyrolysis-incineration Process for Solid Waste Disposal." Battelle Pacific Northwest Laboratories, Richland, Washington, December.

Hokanson, A. E., and Rowell, R. M. (1977). USDA Forest Service General Technical Report FPL 12, 20 pp.

Moddox, R. N., ed. (1975). *Energy Communications* 1(5), 433-494.

National Academy of Sciences (1976). Committee on Renewable Resources for Industrial Materials, "The Potential of Lignocellulosic Materials for the Production of Chemicals, Fuels, and Energy." Washington, D.C.

Reed, T. B., and Lerner, R. M. (1973). *Science 182*, 1299-1304.

State of Maine (1975). "Collected Working Papers on the Production of Synthetic Fuel from Wood," March 31.

Wahlgren, G. and Ellis, T. (1978). *Tappi*, in press.

Wentworth, T. O. (1977). *Chemical Week*, December 21, 38-39.

Progress in Biomass Conversion, Volume I

A SURVEY OF U. S. AND EUROPEAN PRACTICES
FOR RECOVERING ENERGY FROM MUNICIPAL WASTE

James G. Abert
Harvey Alter

National Center for Resource Recovery, Inc.
Washington, D.C.

I.	INTRODUCTION	146
II.	RESOURCE RECOVERY ACTIVITIES	148
III.	FUELS AND ENERGY	150
IV.	MASS BURNING	154
V.	REFUSE-DERIVED FUEL	155
	A. RDF as the Total Fuel	155
	B. RDF as a Supplemental Fuel	160
	C. Co-Firing of RDF in Utility Boilers	165
	D. Co-Firing of RDF in Industrial Boilers	167
VI.	ENVIRONMENTAL CONSIDERATIONS	174
	A. Ash and Emissions	174
	B. Control Regulations	177
	C. Air Emissions Standards for Combined Firing	177
	D. Mass Burning	181
	E. Suspension Firing with RDF with Coal	181
	F. RDF, d-RDF, and Stoker Firing with Coal	185
VII.	CORROSION	188
VIII.	ECONOMICS	190
IX.	RESEARCH NEEDS	195
	APPENDIX	206

I. INTRODUCTION

Municipal solid waste was first used as a fuel to generate steam and electricity in the latter part of the nineteenth century. The first plant in the United States was built in the late 1890s, copying European plants already in operation (Fenton, 1975). In following years, other incinerator plants were built which raised steam for internal power generation; some plants merely incinerated the waste as a means of disposal. Incineration was a significant method of waste disposal in major cities in the United States until it was priced out of the market by low-cost landfills.

In the early 1960s, the concern for clean air was the driving force that raised the cost and closed most of the incinerators in favor of less expensive sanitary landfills. However, the oil embargoes of 1973-1974 renewed interest in using municipal solid waste as a source of fuel. The interest has been reinforced by the constant quest of municipalities for an alternative to disposal, for not having to secure new sites for landfills every so many years.

Principally, the emphasis on using waste as a fuel has been on the historical method of directly burning (incinerating) the combustible portion and recovering the heat as steam. However, there is also interest and developmental activities in mechanical processing to improved solid fuels and chemical decomposition (pyrolysis) and biological digestion to create fuel gases. Some researchers have sought to chemically react waste to form new fuels.[1] The use of waste as a fuel is an alternative to disposal which may reduce the environmental insults possible with traditional forms. Municipal solid waste is a renewable (in a sense) and essentially noninterruptable fuel source.

With the growth of population and, at least during the 1960s and early 1970s a reported increase in the per caput

---

[1] *The U.S. Bureau of Mines has converted municipal solid waste, sludge, and other wastes to a heavy oil and a bitumen on a pilot scale by reacting the waste at high temperature, under pressure, with carbon monoxide and hydrogen, in the presence of alkali as a catalyst. See Appell, et al. (1971).*

waste generation,[1] the potential for waste-based energy could be as much as 2×10^{15} Btu (2 quads) by the year 2000, according to estimates by the Department of Energy. Recovery of the combustible portion of the solid wastes in just the urban areas of the United States in 1974 would have provided about 1 quad (Alter, 1977). At present, it is estimated that in-place waste to energy recovery facilities, those under construction, and those for which there is a firm contract to build, will process 6% or so of the nation's municipal solid waste for energy recovery by 1985. Linear extrapolation of this estimate predicts that 34% of the waste will be processed by 2000, which is relatively small, given the potential of achieving what might be viewed as an energy dividend from this hitherto virtually ignored source.[2] Worse, a nonlinear estimation model is even more pessimistic in predicting future processing capacity (Alter, 1977). Of course, implementation may be accelerated. This is the hope of those who advocate a more purposeful attempt on the part of both public and private sectors to realize the energy potential of mixed municipal wastes.

This chapter first summarizes the current status of energy recovery from municipal waste, both in the United States and abroad. It then looks specifically at a number of the technologies in place or advocated, with a particular

---

[1] *The rate at which the per caput waste generation may be increasing is not well understood. Various reports (Office of Solid Waste, 1973, etc.) have predicted large increases over the recent past and future. However, the actual figures, based on national averages computed by input-output techniques, and reported in the same sources, show a steady rate of per caput generation, if not a small decline. A recent comparison of waste generation in England, for 1931 and 1976, showed only a 12% increase in the mass of solid waste generated per caput over this extended time period. The corresponding increase in volume was 50% (Flintoff, 1978). In addition, there are field reports from disposal facility operators, sometimes substantiated by records, that the generation and increases previously predicted by EPA have not occurred.*

[2] *These percentages are based on U.S. population forecasts of 228 million for 1985 and 250 million for 2000. Total annual waste generation is estimated at 154 million (10^6) tons for 1985 and 169 million tons for 2000. This is a 3.7 lb per caput per day, lower than EPA forecasts for the same time periods (Office of Solid Waste, 1973).*

emphasis on environmental effects. This is followed by a brief section on economics. Finally, the chapter details a number of research needs which have been offered by those working in this area.

II. RESOURCE RECOVERY ACTIVITIES

The Appendix details the status of resource recovery in the United States at the beginning of 1979. Most of the activities listed understandably focus on energy recovery which also disposes of the bulk of the weight and volume of the waste. The principal energy product is steam, produced either by direct burning of the waste or a processed fraction of the waste—what has become known as refuse-derived fuel (RDF).[1] RDF is either burned as the total fuel in boilers especially constructed for this purpose or as a supplement to coal in boilers that were adapted to accept this new form of fuel. As will be discussed, there is not one, but several forms of RDF.

There are several projects intended to produce a gaseous or liquid fuel from the waste, thus broadening market utility. These processes generally employ pyrolysis in some form, although controlled anaerobic digestion of the waste to a fuel gas is being explored as well (Pfeffer, 1976; Kispert, et al., 1976; Wise, et al., 1978). Also, in a number of locations, gas is directly recovered from landfills where its natural production hitherto had been seen principally as a safety and environmental hazard, rather than as an energy opportunity.

Overseas, the main form of energy recovery has been mass burning of the refuse, as it is received. There are a few plants coming on-line which mechanically process the waste to produce a prepared fuel. To list all of the European plants in a form corresponding to the table in the Appendix would be an extensive undertaking. There are currently 243 separate units in Europe to convert waste to energy either in operation, under construction, or planned for the near future (Resource Planning Associates, 1977) installed in approximately 70 separate plants (Engdahl, 1976). Figure 1 illustrates that energy recovery from waste is practiced in virtually every European country. Luxembourg, Denmark, Switzerland,

---

[1]*Other names, such as waste-derived fuel, have also been proposed. However, RDF is the term in common use in the U.S. It was the first proposed by J. Collins (Sheng and Alter, 1975).*

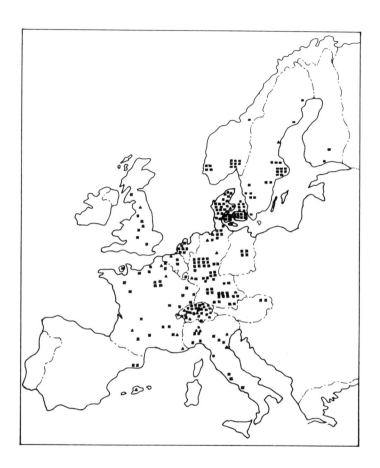

FIGURE 1. Location of waste-to-energy plants (steam generating incinerators) in Europe; ■ in operation, ▲ under construction or planned.
Source: Resource Planning Associates (1977).

The Netherlands, Sweden, and West Germany predominate. It is reported that each has the capacity to recover energy from the waste of more than 40% of its population (Resource Planning Associates, 1977). A separate report indicates that 22% of the waste in Germany is so burned (Umweltbundesamt, 1978) and unpublished reports are that the corresponding figures for Sweden and Switzerland are 30% and 60%, respectively. The energy product is steam for district heating or process use, electricity, cogeneration, or sludge disposal.

It is important to note recent European interest in materials recovery and RDF as alternatives to mass burning. Some European countries are saying incinerator units are becoming too expensive and/or the air pollution control requirements too severe. For a discussion of energy trends and a description of the first RDF plants in Europe, see Alter and Dunn (1979).

Three principal factors encourage the use of waste-to-energy systems in Western Europe. First is climate, which is characterized by a long cold season; second is the prevalence of district heating; there is a shortage of land, particularly in some countries, such as Denmark and Switzerland. While portions of the United States duplicate the climatic conditions of Europe, it is rare to find opportunities for district heating. In general, when they occur in the U.S., they are most likely to involve educational institutions or government buildings rather than residential living accommodations. There are only a relatively small number of these institutions, in contrast to the large potential which residential heating could have to supply markets for waste-to-energy systems.

One U. S. example of using waste for district heating and air conditioning, Nashville was started when downtown urban renewal afforded the opportunity to install a district steam and chilled water distribution system where none had existed before. Thus, a market was created for the products of a waste-to-energy facility of the type appropriate to the need.

III. FUELS AND ENERGY

In the U.S., approximately 80% of residential discards are combustible, assuming no source separation of paper fibers. On a dry basis, waste is about 50% organic materials consisting of: 10% newspaper, 20% other paper, 20% other miscellaneous organic materials (including plastics), and 25% inorganic materials. The latter include about 6% magnetic metals, 1% other metals, 10% glass, and 8% stones, rocks and other inorganic matter. On average, the waste contains about 25% moisture. The heating value in the as-received state may range from 4000 to 4500 Btu/lb.

It is possible to upgrade the quality of waste as a fuel. However, while the quality of the fuel is increased, this is not without cost, both in terms of loss of potential energy as a result of the processing and in terms of the capital and operating expenses associated with the processing itself. It is difficult to say mass burning or some forms of prepared fuel realize significantly more or less of the net energy potential of the waste (Alter, 1977a; Hecklinger, 1976).

Offsetting factors derived from processing vary according to the type of processing, but include: (1) a mitigation of some of the technical problems associated with mass burning; (2) a broader adaptability of the waste-based fuel product to existing units burning coal and maybe oil or gas; (3) an increased ability to transport and store the fuel product; and (4) the ability to extract more of the combustion energy as steam, i.e., a higher thermal efficiency (Hecklinger, 1976).

The output of a waste-to-energy plant, in converting some tonnage of waste to electric power, may be estimated from the nomograph of Figure 2. This diagram also illustrates some of the important variables: output, fuel tonnage, plant heat rate (i.e., Btu/KWH), and load factor. Even this is a simplification but can serve as a first level example or planning tool.

There are various forms of refuse-derived fuel. The text of this chapter discusses several types with the principal emphasis on "fluff," "densified" (d-RDF), and "dust" refuse-derived fuels. In the first and second instances, approximately 60% of the waste is processed as the fuel fraction, and somewhat less in the third. The calorific value of such fuel is of the order of 5000 to 7000 Btu/lb. The "dust" category of refuse-derived fuel is associated with a proprietary process of Combustion Equipment Associates. Processing steps, both mechanical and chemical, produce a finely ground product with a reported heating value of 8000 Btu/lb for laboratory or pilot plant produced material (Beningson, et al. 1975). Note should also be taken of the wet processing approach of the Black-Clawson Company, wherein size reduction and separation are accomplished through wet pulping in machines similar to what are used to pulp wood chips in papermaking. The fuel is what remains after the bulk of the inerts have been removed. This contains approximately 50% moisture so must be burned in bark-burning type boilers. Two current municipal waste-to-energy projects use this approach (Hempstead, N.Y., and Dade County, Florida) to generate electricity (see table in Appendix) (Benziger, et al., 1976).

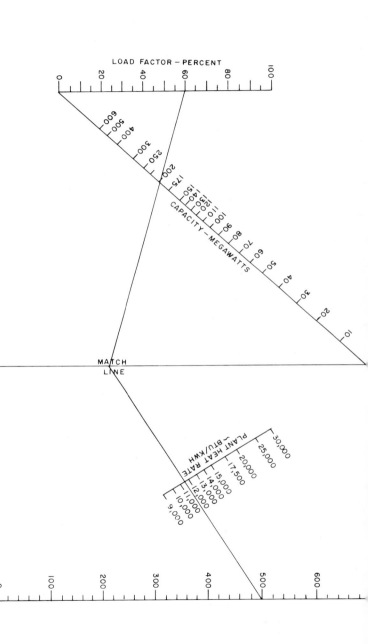

FIGURE 2. Solid waste combustion and generating plant parameters. The nomograph shows the relationship among the capacity, load factor, and heat rate of a boiler and the amount (tons) of solid waste that can be burned in the boiler each day. To use the nomograph, line is drawn (as shown) between load factor and capacity to the match line. A second line is drawn from the match line through the heat rate to indicate the tonnage of waste. Source: Garbe and Levy (1976).

Other possible fuel products are liquid fuels, "medium Btu gas," and "low Btu gas." The main example of the former is pyrolysis oil scheduled to be produced at the Occidental Research Corp., EPA-supported, San Diego County project. It is currently in a state of suspense due to technical and financial difficulties (see Appendix). The product, as produced in an earlier pilot plant and briefly at the San Diego prototype, is not a hydrocarbon oil but rather a viscous liquid degradation product of cellulose. It was intended to be sold to an electric utility as a supplement to residual fuel oil. It has a lower sulfur content than most residual oil, but only 77% of its calorific value.[1] Like high paraffinic oil stocks, it would have to be heated ($\approx 160°F$) to be pumped. It contains water and organic acid moieties and is corrosive to certain types of steel. Although this type of pyrolytic oil is not miscible with hydrocarbon-based residual oil, it can be dispersed and the mixture burned. The continuation of the San Diego project is extremely doubtful, its cost having risen beyond any likely near-term economic application.

Standard pipeline quality natural gas consists most of methane and has a heating value of approximately 1000 Btu per standard cubic foot (Btu/scf). For gas produced from waste, probably anything more than half this calorific value should be considered as a "high-Btu" gas; gas with calorific values about one-quarter or so the value of natural gas labeled as "medium-Btu" gas. Below that, the gas would be identified as "low-Btu." The principal contender for economic production of a medium-Btu gas is the Union Carbide Purox (TM) System which produces a gas with a reported calorific value in the range of 300 to 400 Btu/scf. The principal components are hydrogen (explosive) and carbon monoxide (toxic), so the gas is only suitable for industrial or utility use.

The gas exiting any pyrolysis process has a high latent heat which makes up much of the energy value of the product, hence should be burned directly after production. While the benefits of gas as a refuse-derived energy form are many, at present it does not appear to be economically competitive with mass burning or the various types of refuse-derived fuel, at least in the U.S. There is a recent review of the technology (Jones and Radding, 1978).

---

[1]*This difference in calorific value is by mass. The pyrolysis product has a higher density than oil so the two fuels have comparable calorific values on a volume basis. Liquid fuels are usually sold, and their use measured, by volume.*

IV. MASS BURNING

As mentioned above, mass burning (or incineration with heat recovery) is the most practiced form of energy recovery from waste, particularly in Europe and Japan. There are two different approaches to the combustion of the waste. One is to put the boiler in the firebox, with the water tubes down near the burning grate, known as the waterwall approach. The other is to burn the waste in a refractory firebox with the boiler located in a second chamber removed from that where combustion takes place. The first design appears to increase the efficiency of energy conversion and reduce some operating difficulties. It is the one generally used today.

Worldwide, there have been a number of technical problems with mass burning, particularly corrosion and erosion. These points are discussed in a later section.

There is a difference between the units found in most European installations and those on-line, coming on-line, or being marketed in the United States. About a third of the total number of units (furnaces) in Europe are smaller than about 5 tph. In contrast, U.S. mass burning combustion units tend to be larger. For example, the Resco plant in Saugus, Massachusetts, uses two furnaces, each rated at about 30 tph. This is larger than most of the European installations. (Notable exceptions are two 50 tph furnaces in Paris and a few 40 tph units in Germany.) For the most part, facilities in Europe use a modular approach to achieve larger plant capacity with multiple installations of 5 to 10 tph units. The European approach, of course, works against achieving economy of scale. This can be important, because in the U.S. such economies are often essential to the validity of resource recovery in contrast to landfill. Landfill is less expensive in the U.S. than in Europe; European incinerator plants can afford higher costs.

A disadvantage of mass burning is that it decreases the potential for recovery of metals and glass. While they can be removed from the incinerator residue, this material has been through the combustion process, increasing its recovery cost and reducing its market value. While some ferrous metals recovered from incinerator residue are sold, at present there are no examples of nonferrous or glass recovery and sale (Jenn and Peters, 1971).[1]

There has been a recent interest in the United States in small (1 to 2 tph) so-called modular incinerator systems with heat recovery. These are generally packaged two-chamber starved air units. In the first chamber, partial oxidation occurs, destroying the organic portion of the waste and yielding combustible gases. These gases, plus entrained solid particles, are burned in the second chamber. An afterburner, using auxilliary fuel may also be used so as to consume potential pollutants. The hot gases from the final combustion stage then flow through appropriate heat exchangers to produce steam, hot water, or hot air. As packaged systems, these units are produced in a factory and shipped to the site for installation. They are virtually the only energy recovery method available for small-scale plants competitive in the U.S.

V. REFUSE-DERIVED FUEL

Refuse-derived fuel can be burned in several ways: at the total fuel (100% RDF) in specially designed combustion systems; in modified existing systems as a supplemental fuel; or in systems designed for the co-firing of RDF with municipal or industrial sludge.

A. *RDF as the Total Fuel*

Broad classes of systems for burning RDF as the total fuel have been described by Meissner (1969) as: (1) Packaged systems. These are made of factory-assembled, shippable modules. The upper value of throughput would be on the order of 20×10^6 Btu/h, or 1.2 tph of "enriched" waste having a calorific value of 8500 Btu/lb. (2) Field erected systems.

---

[1]*In the early 1970s, the U.S. Bureau of Mines did considerable analytical work, as well as some pilot testing of processes to recover metal and glass from incinerator residues (Jenn and Peters, 1971). Also, this approach has been researched in several other countries, such as England, France, Denmark, Spain, and Germany. The first commercial plant for processing incinerator residue was scheduled to come on-line in Holland early 1979. It is based on an English design (Alter and Dunn, 1979).*

These are larger, specially designed systems erected on the site; they are more costly, but better adapted to the local conditions, thus expected to be more reliable. (3) Fluidized bed systems. These might conceivably belong to either class above, but will be discussed separately, since most are experimental technology.

Packaged systems are generally of the starved-air, rotary kiln, augered-bed, or basket-grate type. In the starved-air type combustor the fuel is often batch-fed and incinerated in less than stoichiometric air. The off-gases are burned to complete combustion in a secondary chamber, sometimes with auxilliary fuel (gas or oil) and injected secondary air. Supplementary fuel may be used in a separate (virtually) after-burner as a pollution control device.

In the rotary kiln, combustion takes place in a rotating cylinder with refractory walls. The noncombustible material (ash) flows down into a hopper at the end of the kiln. An auxilliary fuel burner may be used as well as an after-burner as a pollution control method. The off-gases can be directed to a waste-heat boiler. There are field reports of rotary kiln incinerators for mixed municipal waste experiencing premature corrosive failure.

The augered-bed combustor is fed continuously by a hopper. The waste is driven along a refractory-lined combustion chamber and burned with excess air. The off-gases can be directed through a boiler bank. Ash removal is done continuously.

The basket-grate system rotates a truncated-cone grate with its axis inclined at an angle of about 30° on the horizontal. After-burning is effected in a secondary chamber on top of the primary chamber.

These various packaged systems were designed originally for unprocessed waste. Although they could conceivably be used for RDF, there are additional important considerations. Because RDF is a processed waste fraction and much of the noncombustible material has been removed, its calorific value and heat release rates have been increased. The preparation of some form of RDF from waste requires mechanical processing, the cost of which dictates an economy of scale usually much higher than the capacity of any of the packaged units. In short, facilities fired with RDF as the total fuel are likely to be large, consuming of the order of 750 tons per day (tpd), or more, of waste. This can be seen from the examples in the Appendix; the several examples of RDF as the total fuel (Akron, Niagara Falls, Albany) are large plants with field-erected boilers.

The field-erected systems may be of the stoker-grate type or semisuspension type, similar to those used for wood waste. At present, only stoker units used for this application are bark-type boilers for the wet RDF produced by the Black-Clawson plants. The semisuspension boilers use a shredded waste (after removal of magnetic metals) or a shredded and air-classified waste as the RDF. The fuel is charged pneumatically and burns in suspension as it falls onto a grate. Combustion is completed on the grate, which is also the means of ash removal. A field-erected boiler for this application is designed to handle the larger volumes of combustion gases and higher ash anticipated with RDF compared to say coal or oil. A brief description of the boilers has been given (Alter and Dunn, 1979).

No intrinsically new or different system has been designed to accept d-RDF or dust RDF as the total fuel, nor is it likely to be required. An advantage of d-RDF is that it can be fed by coal stokers with little or no modification. An advantage of dust RDF is that it can be used in pulverized coal boilers, or possibly suspended in oil, with little modification.

Fluidized bed combustion of RDF and d-RDF has been investigated in the U.S. at a few pilot plant installations. In one such system at Combustion Power Company, a shredded and air-classified light fraction was combusted using a bed of sand as the fluidizing medium (Chapman and Wocasek, 1974; Chapman, 1975). Combustion temperatures were in the range of 1500 to 1800°F. The combustion gases, after cleanup of entrained particulates, fed a gas turbine and electrical generator assembly. A feasibility project at Stanford University proposed using a fluidized bed of sand four feet thick to burn raw refuse to produce steam for heating and electricity generation on the campus. In this scheme, the problems of running dusty gases through a turbine are avoided (Mean, 1977).

The two main products derived from refuse burning are steam and electricity. Figures 3 and 4 depict the steps in realizing these outputs. There are a number of such projects underway or in the developmental stage. First, the Akron, Ohio, 1000 tpd "Recycle Energy System" is presently under construction and is scheduled to be fully operational by December 1979. It will use RDF as the sole fuel in a semisuspension fired, field-erected boiler somewhat similar to one used in Hamilton, Ontario, for shredded waste (after removal of magnetic metals) and to those used for burning wood wastes.

It is said that "the elements of the process are all essentially off-the-shelf technology" (Glaus, et al., 1974). Up to 1350 tons of raw waste can be stored. Two 60 tph shredders will handle the nominal 1000 tpd in slightly more than one eight-hour shift. After air classification, the light fraction will be stored in an 1860-ton traveling screw,

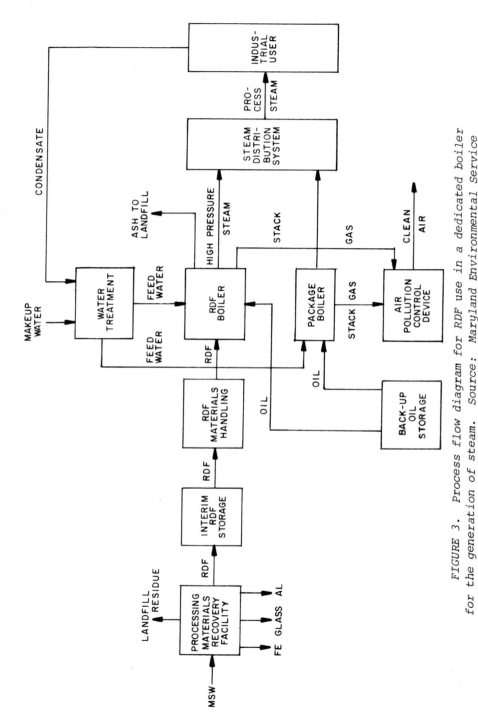

FIGURE 3. Process flow diagram for RDF use in a dedicated boiler for the generation of steam. Source: Maryland Environmental Service (1978).

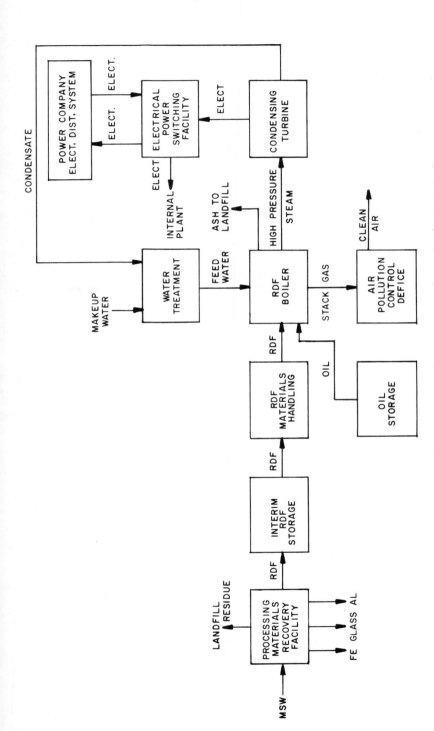

FIGURE 4. Process flow diagram for RDF use in a dedicated boiler for the generation of electricity. Source: Maryland Environmental Service (1978).

self-unloading bin, followed by a pneumatic conveying system carrying the RDF to a surge hopper and the boiler.

More than 70% of the in-house power requirements will be provided by turbines, which also reduce the header pressure steam for feedwater heating and de-aerating requirements. The three boilers (125,000 lbs/h of steam each) are equipped with grates for burnout of the larger particles, but a significant part of the combustion will take place in suspension. Rated steam outlet conditions are 560 psig, 579°F (saturated). The design is for a heat release of 750,000 Btu/ft^2 of grate. The geometry of the boiler and economizer surfaces is designed to minimize bridging and plugging.

The electrostatic precipitators are designed for particulate emissions of 0.11 lb/10^6 Btu, or half the allowable state level. A regenerative air heater follows the precipitator. The boiler is expected to have a 78% efficiency at maximum load and a gas exit temperature of 325°F.

The customers for steam will be about 200 accounts of a district steam system purchased from Ohio Edison, including two hospitals, a governmental complex, and major office and commercial buildings. In addition, a local industry, B. F. Goodrich, will buy excess steam on an interruptible basis.

Other projects using RDF as the sole fuel are the Black-Clawson plants mentioned earlier. Moisture levels of 50% would be a problem in most industrial boilers and it might make sense to use part of the fuel to dry the bulk of it. However, drying involves a loss of net energy which can reach 8% for a 10-point reduction in moisture content (Systems Technology, 1975). The energy cost for drying RDF has been discussed (Sheng and Alter, 1975). Some boilers have been designed to fire "difficult" fuels such as agricultural waste and forest residues. These fuels are also high in moisture content and the design of these boilers are such as used for hogged fuel or bark.

In Hempstead, New York, the 2000 tpd plant uses a "Babcock France" bark-type boiler to generate 600 psig, 750°F steam. The steam is sold to Long Island Lighting Company to run two 20 MW turbogenerators located on the premises. The 3000 tpd facility to be built in Dade County, Florida, is designed along the same principles.

B. *RDF as a Supplemental Fuel*

1. *Co-Firing with Coal.* An early experiment in firing fluff RDF as a supplemental fuel to pulverized coal took place in 1972 at the Meramec plant of the Union Electric Company in St. Louis, Missouri (Gorman, *et al.*, 1977). A joint venture of the U.S. Environmental Protection Agency,

the City of St. Louis, and the Union Electric Company, it was a 300 to 400 tpd demonstration project. RDF was burned in existing, tangentially fired pulverized coal units. Four modified nozzles were installed on top of each of the four corner burner assemblies. A pneumatic conveying system, with blowers, feeders, and piping, was also part of the retrofit. A similar project, in Ames, Iowa, has also been described in some detail (Even, et al., 1977).

There has also been some experimental work with densified-RDF, called d-RDF. This fuel is being developed for cocombustion with coal in stoker-fired boilers, generally in the range of 50,000 to 200,000 lbs/hr of steam. An advantage of the d-RDF, which is produced in the same size range as the coal used, is that it can be fed to the combustion chamber along with the coal, thus eliminating the need for separate firing systems as for fluff RDF use. Being sold to smaller purchasers of coal than the larger users associated with the fluff RDF market, the selling price for d-RDF would be expected to be higher. The differences between the price paid for fuel of large coal buyers, such as the electric utilities, and smaller industrial or institutional users such as would use d-RDF, and the lower cost of storing and feeding the d-RDF, are seen as being sufficient to cover the cost of the densifying step.

A test burn using d-RDF pellets was conducted at Wright-Patterson Air Force Base in 1975. Two d-RDF/coal mixtures were tried: 40:60 and 58:42 by weight, or 1:1 and 2:1 by volume, of 23:77 and 37:63 by calorific value. The 1:1 mix firing lasted a total of 34 hours, with one nonstop firing of 15 hours; the 2:1 mix firing was 6 hours, nonstop (Buonicore and Waltz, 1975).

The boiler used in the firing test was of the forward traveling grate stoker type. An overthrow motor distributed the fuel across the traveling chain grate. The boiler was rated at 80,000 lbs/h (125 psig saturated steam); normal load at the time of the test (summer) was 25,000 to 30,000 lbs/hr. The main conclusions from these tests were: (1) No significant problems appeared in the 1:1 RDF/coal mix firing. (2) The use of a 2:1 RDF/coal mix caused operating problems and might require modifications to the boiler for steady operation, such as preventing an uneven distribution of the fuel (and the fire) on the grate and a loss of load control. (3) Corrosion and refractory deposits did not appear to be a matter of concern. (4) Only a nominal effect on ash handling was noticed at a 1:1 mix.

There have been other brief experimental burns with d-RDF. In 1971, an alfalfa cubetter was used to densify shredded industrial waste in Fort Wayne, Indiana. These "cubettes" were fired with coal in a stoker boiler with apparent success but the project was not continued, mainly because the boiler was in need of a complete rehabilitation and addition of more sophisticated air pollution control equipment, both unrelated to the burning of "cubettes" (Hollander and Cunningham, 1971; Weston, 1971). Also in the early 1970s, Sira Corporation (now Vista Fiber and Chemical Products, Inc.) began making densified fuel which has been the feedstock for a series of small test burns. Tests were conducted at Eugene, Oregon, and Oshkosh, Wisconsin.

The U.S. Environmental Protection Agency has supported two progressively larger test burns. The first of these was conducted at the Maryland Men's Correctional Institution in Hagerstown, Maryland, in 1977. The second will take place in 1979 at the General Electric Company plant in Erie, Pennsylvania. Properties of the fuel mixtures burned in the first test are given in Table I. Overall, in the three tests involving d-RDF, more than 300 tons of the material were burned. Tests were one to two days in length. The general conclusion was that it was usually difficult to discern a difference between the coal only and coal/d-RDF firings. Because of the lower sulfur content of the d-RDF, the SO_2 emission decreased as more d-RDF was substituted for coal. Conversely, the chlorine emission increased. The substitution of d-RDF for coal caused a more violent and volatile fireball in the furnace which appeared to lead to better burnout and the elimination of smoke. Opacity of the stack emissions decreased as larger proportions of d-RDF were used. (The boiler was not equipped with an electrostatic precipitator.) The results led to the decision to undertake the large, longer tests at General Electric which will involve a boiler producing superheated steam.

Similar tests of the use of d-RDF have been conducted in England where d-RDF pellets of 9, 15 or 19 mm diameter, by about 25 mm long, were burned alone or with coal in a Parkinson Cowan "Vekos Powermaster" multifuel package boiler. When operating on d-RDF as the sole fuel, boiler efficiencies in the range of from 72 to 86% were obtained at up to 99% of the boiler rating. The boiler manufacturer estimated that about 80% thermal efficiency is normally attained when firing coal. Gaseous emissions ranged 100 to 260 ppm NO_x with 50 ppm or less of hydrocarbons; chloride emissions were 141 and 602 mg/m^3 for two measurements. This type of boiler has manual ash removal and the higher ash content of the RDF, compared to coal, resulted in an unacceptable rate of buildup of ash on the fixed grate. However, this is a problem peculiar to the type of boiler. No such problems were observed when coal:d-RDF ratios of 1:1 were used (Birch and Jackson, 1979).

TABLE I. Test Burn Densified Refuse-Derived Fuel (d-RDF)

	Blend of coal to d-RDF by volume (as received) Volume Ratios			
	1:0	1:1	1:2	0:1
Moisture	1.3	6.6	7.9	16.6
Volatiles	22.6	31.7	38.3	48.6
Fixed carbon	54.2	38.4	29.1	9.0
Ash	22.0	23.3	24.7	25.9
Calorific value	11,607	8,988	8,392	5,103
Carbon	66.5	54.1	47.3	30.9
Hydrogen	4.3	4.1	4.1	3.8
Oxygen	3.4	9.8	14.2	21.8
Nitrogen	1.3	1.1	0.9	0.6
Sulfur	1.2	0.86	0.66	0.23
Chlorine	0.05	0.15	0.21	0.33
Fusion temp., °C				
Init. def.	1,273			1,116
1st soft.	1,308			1,151
2nd soft.	1,335			1,179
Fluid	1,371			1,213
Theoretical air				
kg/kg of fuel	9.04	7.26	6.27	3.93
% weight rate d-RDF	0	35	52	100
% heat rate d-RDF	0	20	37	100

Note: Unless otherwise noted, all values are weight percent.
Source: Rigo, et al. (1978).

Combustions trials of several hours duration were also conducted in England on several industrial chain grate stokers. The results showed that the smaller sizes of d-RDF could be burned satisfactorily, if not alone, then in 1:1 mixture with coal. The thermal efficiencies and percentages of maximum boiler ratings achieved are reported reasonable in view of the limited amount of fuel available for the tests and the lack of operating experience with the new fuel (Birch and Jackson, 1979).

2. *Co-Firing with Oil or Gas*. Oil and gas are fuels with low ash content, typically 0.1% for residual fuel oil and substantially zero for natural gas. Furthermore, the

excess air requirements for complete combustion range from 5 to 7% (or lower). These differ from those required for burning RDF only, possibly in excess of 40%.

Some literature indicates that dust RDF would behave as pulverized coal. Mixed fuel burners (coal, oil, or gas) could be fed with dust RDF as a replacement for coal provided the boiler was fitted with ash handling equipment. Combustion Equipment Associates, the developer of Eco-Fuel II®, a dust RDF, has conducted firing tests of slurries of this fuel with residual oil at up to 40% RDF by weight. The mixtures were found stable over long periods of time. According to the company, . . . "firing through conventional oil guns with modified nozzles and diffusers, the combustion envelope is shorter than with oil alone, free of sparklers, and the stack is smoke-free" (Beningson, et al., 1975). A study was conducted to identify the retrofit requirements for co-firing of oil and RDF in mediumsized institutional boilers [up to 70,000 lbs/h of steam (Renard and Fretz, 1979)].

3. Coincineration of RDF with Sludge. Sludge, the product from a waste-activated sewage treatment and/or industrial organic material contains incinerable organic components, with varying amounts of water and dissolved salts (Novak, et al., 1977). After chemical conditioning and mechanical dewatering, a wet cake is obtained, containing 70% or more water. Its ash content is 15 to 20%, approximately 75% of which results from conditioning chemicals.

Most sludge incinerators are of the multiple-hearth type (Sieger and Moroney, 1977). The combustion and cooling air enter the lower part of the incinerator and flow upward, counter to the descending flow of sludge. From the bottom up, the zones can be labeled as cooling, combustion, and drying. There are also rotary hearth (cyclonic) and fluidized bed installations. For a review, see Niessen, et al., (1976).

Coincineration of sludge and RDF was investigated in a pilot installation in Contra Costa County, California. The RDF was a shredded, air classified, screened light fraction from mixed municipal waste (Bracken, et al., 1976). The sludge had a solids content of 16%, a volatile solids content of 75%, and a calorific value of 9000 Btu/lb of dry solids.

Incineration and pyrolysis were the two operating modes investigated. In incineration, self-sustaining combustion of a sludge having a 16% solids content was possible at a 1:2 RDF:sludge ratio. Sludge must contain 24% solids for autogenous combustion without addition of RDF. The addition of RDF enhanced heat recovery and allowed the reduction of sludges with high water content without supplemental fuel consumption.

In Franklin, Ohio, the wet solid waste fraction from the Black-Clawson pulping process was mixed with sludge from a water treatment plant, mechanically dewatered to 45% solids content, and injected into a fluidized bed furnace (Babcock and Wilcox, 1973).

C. Co-Firing of RDF in Utility Boilers

Springboard for the development of the co-firing was the EPA-sponsored demonstration of coal and RDF firing at the Union Electric Company's Meramec power plant in St. Louis. The RDF was produced by the City. The demonstration covered the years 1972 to 1975, during which time some 40,000 tons of RDF were burned. Several other projects resulted from the success of this demonstration including plants in Milwaukee, Wis.; Chicago, Ill.; Monroe County, N.Y.; and Ames, Iowa.

The Milwaukee resource recovery plant, owned and operated by the Americology Division of American Can Company, has a design throughput of 1600 tpd (Duckett, 1977). Early 1979, it was processing about 900 tpd. The RDF produced results from a primary shredding (6-10 inch nominal), air classification, and shredding of the light fraction (to 95% less than 3/4 inch). This fluff RDF undergoes further magnetic separation and screening. It is shipped to the Oak Creek Plant of the Wisconsin Electric Power Company, where it is stored in a 650 ton capacity Atlas bin. From the bin, the fuel is conveyed pneumatically into the boiler through separate nozzles above the pulverized coal feed points. The fuel, after screening, should have an ash content of less than 20% and a calorific value in excess of 5000 Btu/lb, as-received. About 50 to 55% of the incoming MSW becomes RDF.

In August 1977, the rate of firing was 10 to 15 tph of RDF. An increase to 40 tph, or about 15% heat input of RDF, is planned. The power company boiler produces 2×10^6 lbs/h of steam at 2620 psig, 1050°F. Some slagging of the tubes, presumably due to the original high ash content of the unscreened fuel (up to 30%), was observed. Rotary screens (disc type) were installed in 1978. No significant increase in fly ash is reported as a result of co-firing, but there is an additional bottom ash loading, with much floating material (wood and plastics) in the sluice.

The Southwest Supplementary Fuel Processing Facility of the City of Chicago is designed to handle 1000 tpd, or about 20% of the city's municipal refuse (Godfrey, 1976). Coarsely shredded material is fed into an 80 tph air classifier. The light fraction is conveyed to a cyclone for separation and from there to a secondary, 60 tph shredder for size reduction to less than 1-1/2 inch. The fluff RDF is transferred to live-bottom bins having a capacity of 1-1/2 days' supply, or 150,000 ft^3.

The fuel will be fired in an existing 222 MW Combustion Engineering twin-furnace, tangentially fired pulverized coal boiler, and in a similar unit rated at 347 MW.

The Monroe Coutny, N.Y. facility, scheduled for startup 1978-1979, has a design capacity of 2000 tpd. From this, 1200 tpd of fluff RDF, suitable for suspension burning in existing coal-fired boilers, will be produced. A description of the Raytheon process has been given (Kenyon, 1975; Carlson, et al., 1976). The raw refuse undergoes primary shredding and air classification, magnetic separation, trommel screening, secondary shredding and air classification. The RDF (fluff) specifications are: "Btu of fuel shall be 5000 Btu/lb or greater, or an as-received basis; material should be free-flowing and injectable into the fuel firing mechanism at all times; free organics should not exceed 10% by weight; the test procedure shall be mutually agreed upon between Rochester Gas and Electric (the fuel purchaser) and the fuel supplier."

Two C. E. tangentially fired, suspension-type pulverized coal boilers will be converted for burning the RDF at up to 10% heat input. Assuming successful initial tests, the two other boilers would also be retrofitted to accommodate RDF. For a minimum of 20% heat input from RDF in these four boilers, the utility will consume 545 tpd of RDF (seven-day week). Markets will be needed for the remaining 600 tpd.

A plant in East Bridgewater, Mass., is producing Eco-Fuel II® from up to 1200 tpd of refuse. Based on this prototype plant, others are under construction in Bridgeport, Conn., and Newark, N.J.

As previously mentioned, Eco-Fuel II is a processed fuel. An objective is to produce a fuel with a low ash (about 7%) and sulfur (0.3 to 0.5%) contents. The product of the experimental trials is claimed to have a heating value of at least 8000 Btu/lb and 2% moisture, equivalent to that of a Montana coal. The particle size of the fuel can be controlled in the final pulverizing step (Industry Trends, 1977). To produce Eco-Fuel II®, the waste is shredded in a flail mill, followed by a separation of the magnetic metals. The material is screened to remove grit and glass. The screen underflow is reported to be 60% ash. The product is treated with an inorganic chemical (e.g., an oxidizing agent or mineral acid) to embrittle the cellulose. Grinding in a ball mill (with hot steel balls at 200-400°F) increases the effect of the chemical agent. A powder of average particle size of 0.15 mm (100 × 80 mesh) can be obtained with a power consumption of 33 KWH/ton instead of 50 to 105 KWH/ton which would be required without the embrittling agent. The process allows for the recovered combustible wastes not suitable for Eco-Fuel II® (unground ball mill product and combustible screen rejects) to be used as the fuel required in the drying process.

According to the developer, pulverized Eco-Fuel II® was burned in a boiler with residual fuel oil and over a range of mixture ratios. Extensive burning of Eco-Fuel II® slurries (up to 40 wt. %) were conducted successfully by the manufacturer. Compaction in briquettes (with a small amount of binding agent) is also possible; the briquettes have been pulverized by conventional coal equipment by a U.S. utility.

In Bridgeport, Conn., CEA plans to sell the fuel produced by its 1900 tpd facility to United Illuminating Co., which will fire it in two Babcock and Wilcox wet bottom boilers rated at 85 MW and 180 MW. One boiler is rated to produce 625,000 lbs/h steam at 1500 psig and 1000°F; the other produces 1.5×10^6 lbs/h at 2000 psig and 1000°F. CEA's Newark plant will produce fuel for the Public Service Electric and Gas Company. The boiler is a Foster Wheeler pulverized coal and gas wet bottom furnace, retrofitted to burn oil. This twin furnace boiler is rated at 1.9×10^6 lbs/h of steam at 2350 psig and 1100°F (Edison Electric Institute, 1977).

D. Co-Firing of RDF in Industrial Boilers

A major unrealized market for the energy potential of municipal refuse may be the direct burning of RDF by industrial energy users (Singer and Mullen, 1974). Many industrial plants which currently burn conventional fuels may have the potential to burn local refuse as a supplemental or primary fuel (Schwieger, 1975).

Potential users include hospital complexes, education, and business facilities which use district heating and air conditioning, and a wide array of industrial installations.

Electric utility companies are most often mentioned as the major potential consumers of solid fuel derived from municipal refuse (National Center for Resource Recovery, 1975). Utility generating stations can use substantially larger amounts of RDF than any one private industrial user. For example, from Figure 2, a power generating boiler, 100-MW capacity and a 80% load factor at a 10,000 Btu/KWH heat rate, will consume 300 tpd of refuse or its equivalent RDF, or the daily generation of approximately 165,000 people.[1] It is likely to take several industrial users to consume an equivalent amount of waste. Thus, if RDF is to be sold to industries, a diverse and decentralized market will result.

Equipment for burning falls into four categories:

(1) Existing boilers designed to operate on conventional fuels, some of which may be able to burn RDF or d-RDF with or without modifications to the boiler and its feed mechanism and/or various degrees of preprocessing of the refuse.

(2) Existing boilers designed to burn various solid fuels such as wood bark and bagasse as supplemental or primary fuels; such boilers should be able to accommodate some form of refuse as a fuel without major modifications.

(3) New conventional boilers with the capability for refuse burning as a design characteristic.

(4) New boilers specifically designed to handle refuse as a "difficult to burn" material.

Any development agency planning to market RDF to public sector facilities and industrial users should conduct a thorough survey of all installations potentially capable of burning refuse in existing boilers. Table II presents the steam requirements and potential RDF use of eight basic categories of potential users. A combination of several of these categories can be found in many communities.

For example, a typical hospital might require 40,000 lbs/h of steam. The annual steam requirement (eight months heating and no air conditioning) would be approximately 80×10^6 lbs/year. At 70% efficiency, this necessitates approximately 4600 tons of 12,500 Btu/lb coal per year. If RDF were used as the sole fuel source to produce the same amount of steam, 9580 tons, with a calorific value of 6000 Btu/lb would be required. This would amount to an average of 1-1/2 transfer trailers of RDF per day (65-75 cubic yard type—19 tons per load).

The amount of d-RDF which might be burned at a small hospital, community college, office building, or canning plant to meet annual demands would be equivalent to one transfer trailer/day or less. Thus communities having enough waste generation to justify producing 250 or 300 tpd of fuel must find several users. The unlikely exception would be an

---

[1]*This estimate is based on 300 days per year operation of the boiler and generation of waste of 3 lbs per caput per 365 days per year. Similar estimates can be made for RDF as a supplementary fuel in fossil-fueled boilers with about the same result.*

existing boiler designed to burn fuels which would take refuse with little or no processing.

Larger installations— lumber mills, paper mills, large industrial processing plants—are better candidates for RDF burning. Many of these plants have boilers built to accommodate wood chips, waste lumber, industrial sludge, and similar wastes and would require less capital expenditure to accept some form of RDF. Table III lists the fuel consumption and steam generation capacity of categories of industrial plants. These plants generally use stoker coal boilers, 100,000-250,000 lbs/h capacity. Many of these plants could convert existing equipment to burn RDF and d-RDF.

Oil and gas fired boilers present problems of no ash handling capacity and, often, little or no air pollution control equipment. Hence, the cost of converting to accept some form of RDF will be high. Prior to 1974, many coal-burning boilers were converted to oil or gas to avoid investment in air pollution control equipment. Boilers of this type may prove to be good candidates for RDF burning. The potential fuel cost saving may offset the cost of air pollution control equipment.

There are significant differences in the economics of RDF use by a typical industrial boiler user (at say 100,000 lbs/h of steam) and a utility with a boiler in the range of 100 MW (approximately 1×10^6 lbs/h of steam). Utilitites amortize conversion costs over a large quantity of fuel. Because coal-burning utilities make large purchases, they generally procure their fuel at a lower cost than smaller industrial users. Bulk purchasing also works to their advantage in the case of other fuels. Industrial users could conceivably pay higher prices than utilities for the same amount of energy. The higher prices that industrial users pay would indicate that energy recovery from MSW could compete in places where it might not if only a utility market were available.

Although alternative fuel costs improve the economics of RDF for industrial users, manufacturers would still require some investment for fuel handling and storage facilities. The question is whether these costs could be compensated by a lower (i.e.g, discounted) fuel cost for RDF compared to the conventional fuel.

As an illustration, consider installation of ancillary equipment to accept RDF costing $1 million and having a useful life of ten years. At an interest rate of 8%, the annual cost of conversion would be $149,000. For a boiler with a 100,000 lb/h capacity, which would be fired with 10% RDF (heat basis), approximately 7300 tons per year of RDF could be consumed. If the RDF had a calorific value of 6000 Btu/lb, the capital recovery is equivalent to $1.70 per million

TABLE II. Steam Requirements and Potential RDF Uses for Private Installations

Plant	Operating time h/y	Steam requirements lbs/h	Steam requirements 10^6 lbs/h	Heat capacity[a] 10^6 Btu/y
Hospital, medium sized	6,000	6–10,000	30–60	34.9 to 58.2
14-story office building	1,500	7,000	10.5	10
Local Community College (Northeastern location)	2,000	30,000	60.0	58.2
Small seasonal canning plant	1,000	3,000 low 50,000 peak (avg 18,670)	19.7	18.1
Medium-sized chemical plant	6,000	40,000	240	234
Large-sized chemical plant	6,000	125,000	750	728
Paper mill	6,000	175,000	1,050	1,019
Lumber mill	6,000	180,000	1,080	1,048

[a] 970 Btu/lb steam.

Source: Compiled from Autem (1975); Carter (1974); and Babcock and Wilcox (1963).

TABLE II. (Continued)

RDF needed tons/y @ 5,000 Btu/lb	Transfer trailers (annually) (18 tons/load)	Trailer deliveries Avg 350 days/y	Comments
3,492 to 5,820	194-323	1 trailer/ 2 days	Use low & high pressure steam for heat, autoclaving, and air cond.
1,010	56	1 trailer/ 6 days	Heat & air conditioning
5,820	323	1 trailer/ day	Three boilers share the load, air cond., heating and some autoclaving.
1,810	101	1 trailer/4 da 1 trailer/da, in season	Seasonal processing operation, four months annually
23,830	1,293	4 trailers/ day	Continuous operation.
72,750	4,042	12 trailers/ day	Continuous.
101,900	5,661	16 trailers/da	Continuous.
10,480	3,822	17 trailers/da	Continuous.

TABLE III. U. S. Total Steam Generation Capacity
(by Industrial Category)

Industrial category	10^6 lbs/h total steam generation	Btu/year $\times 10^{12}$		
		Coal	Oil	Natural gas
Petroleum and coal products	7	1380	9	88
Primary metals	1.7	542	233	1150
Chemical and allied products	6	462	213	1490
Paper and allied products	5.5	236	407	498
Good and associated products	4.3	115	131	57
Lumber and wood	2.3	5	40	34
Transportation and equipment	3.6	65	43	148
Miscellaneous	5	3	20	42

Source: U.S. Environmental Protection Agency (1975).

Btu. Therefore, even if the RDF were free, the alternative fuel must cost more than $1.70/$10^6$ Btu. Figure 5 depicts the relationships involved. This graph allows for an approximation of the capital recovery cost associated with various boiler adaptation costs, boiler size, and the percent of RDF to be substituted. Use is clockwise, beginning on the left side, horizontal axis (Abert, 1975).

For small boiler installations, economics dictate a much higher burning ratio (RDF to conventional fuel). However, the use of higher percentages of RDF in industrial boilers has not yet been extensively tested and, therefore, boiler manufacturers are so far reluctant to endorse the practice.

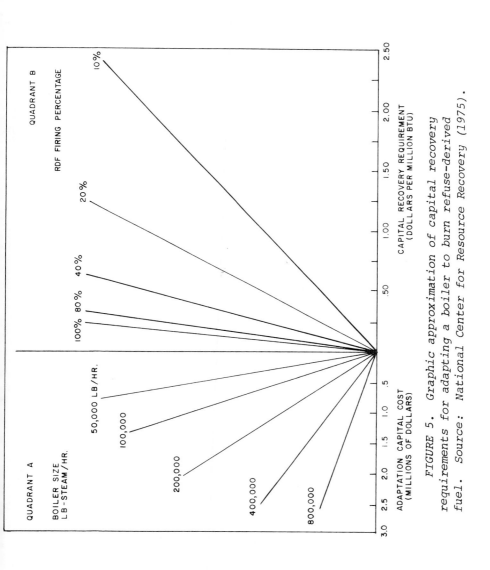

FIGURE 5. Graphic approximation of capital recovery requirements for adapting a boiler to burn refuse-derived fuel. Source: National Center for Resource Recovery (1975).

VI. ENVIRONMENTAL CONSIDERATIONS

A. Ash and Emissions

Some MSW is noncombustible, some components have an inherent ash (e.g., the clay in paper coatings or naturally ash occurring in wood). During burning, part of the ash rises with the hot gases from combustion ("fly ash") and part stays on the grate or falls to the bottom of the combustion chamber ("bottom ash"). The amount of ash is also a function of the type and efficiency of the combustion process.

Dry unprocessed waste burned in the laboratory has a residue of about 29% ash, some three-fourths of which is metal and glass. This is detailed in Table IV. Most of the fly ash comes from the remaining one-quarter. Using the data of Table IV, for every ton of waste fired—including its contained moisture—110 lbs of ash will be generated, not from metals and glass, maximum. This compares to an overall ash content, including the metals and glass, of 416 lbs per ton (Niessen and Chansky, 1970). (All of these figures are averages for MSW as-received, and not for any form of RDF.)

Five factors are the chief determinants of the gross amount of particulates or fly ash likely to result during combustion: (1) the refuse composition, (2) the completeness of combustion, (3) the burning rate, (4) the grate system, and (5) the underfire air rate. Generally 10 to 20% of the fine ash will be emitted as particulates by mass burning incinerators (Niessen and Sarofim, 1968). This yields 11 to 22 lbs of particulate matter/ton of refuse burned. This gas-suspended particulate matter must be treated to meet local air pollution control regulations.

Poor combustion of the waste can result in emission of products other than particulate matter, e.g., carbon monoxide, hydrocarbons, oxygenated hydrocarbons, and other complex cyclic compounds. No specific point source emission regulations yet exist and good operating procedures should minimize their production. Various quantities of inorganic gases such as sulfur and nitrogen oxides (SO_x and NO_x) may also be emitted. The emission of NO_x may result both from the nitrogen content of the waste and the oxidation of nitrogen in combustion. No specific federal emission regulations set point source standards for waste incineration for these gases, although NO_x and SO_x emissions from power plants burning conventional fuel are regulated (see Table V) and ambient air quality standards exist for most of these gaseous emissions.

Table V also gives the particulate standards for fossil fueled boilers; 0.10 lbs/10^6 Btu is approximately the same

TABLE IV. Refuse Composition (100 lbs of Dry Solids)

	% Moisture as received	Composition wgt % (dry basis)	Ash lbs (dry basis)
Metal	6.6	11.19	10.13
Paper	24.3	45.59	2.74
Plastics	13.8	1.50	0.17
Leather and rubber	13.8	2.05	0.24
Textiles	23.8	2.38	0.08
Wood	15.4	2.96	0.09
Food waste	63.6	9.90	2.17
Yard waste	37.9	10.79	0.54
Glass	3.0	11.32	11.21
Miscellaneous	3.0	2.32	1.62
Total	28.3[a]	100.00	28.99

[a] Weighted average.
Source: Moisture column: W. R. Niessen and S. H. Chansky (1970); weight of ash: Niessen and Sarofim (1968).

TABLE V. Federal Point Source Performance Standards for Fossil Fuel-Fired Boilers

	SO_2		NO_x		Particulate
	$lb/10^6$ Btu	ppm	$lb/10^6$ Btu	ppm	$lb/10^6$ Btu
Coal	1.20	520	0.70	525	0.10
Oil	0.80	550	0.30	227	0.10
Gas			0.20	165	

Source: Bogot and Sherrill (1976).

as 0.06 grains/scf (Bogot and Sherrill, 1976).[1] Presumably, the same point source standards would apply to power plants burning RDF with fossil fuel. An exception allowing a modification to burn RDF as a supplementary fuel, without reclassifying an old boiler as a "new source," has already been made.

The NO_x, SO_x and hydrogen chloride (HCℓ) emissions from incineration of solid waste have been reported by several. One example (Kaiser and Carotti, 1971) is listed below, measured at the furnace exit.

NO_x	23 - 25 ppm
SO_x	33 - 40
HCℓ	455 - 732

Solid waste has a low sulfur content, about 0.2% by weight. Most coals and residual oils used today range from about 1 to 3% sulfur (U.S. Environmental Protection Agency, 1973). Further, both the incineration ash and wall slags contain sulfur compounds, thus some sulfur is not released as oxides in the stack. Thus, sulfur oxide emissions from solid waste burning generally are below the most stringent present or anticipated restrictions.

Restrictions on NO_x emissions from fossil fuel combustion have been formulated. NO_x emissions are greater from such fuels than from MSW. It appears unlikely that this pollutant will pose a problem in burning 100% MSW, or RDF, nor increase emissions when substituted for coal.

From a health perspective, the composition of the emissions is at least as important as the total concentration. Several studies of the trace metal concentrations of boiler stack emissions during combustion of RDF have been made. Compared to firing of coal alone, combustion of RDF results in somewhat increased emissions of some trace metals (e.g., lead) (Gorman, et al., 1977; Jackson, 1976; Even, et al., 1977; Shanks, et al., 1977; Greenberg, et al., 1978).

It is difficult to assess the health significance of increased lead emissions. However, the following case can be considered as a rough indicator from co-firing on coal/RDF mixture. The average reported concentration of lead in precipitator controlled emissions where RDF supplied 10% of the heat input in a utility boiler was 343 µg/m^3 (Gorman,

---

[1]Hence, 1.67 times the particulate discharge measured in grains gives an estimate in units of lbs/10^6 Btu. For stoker-fired units, the factor is about 2.2 for coal of 12,000 Btu/lb burned at 50% excess air.

et al., 1977). Applying a conservative air dispersion or dilution factor of 1/1,000 to this level (Stern, et al., 1973), leads to a conclusion that the highest possible ground level concentration of lead attributable to the power plant would be 0.34 µg/m^3, about 20% of the EPA standard (1.5 µg/m^3).

The effect or RDF combustion on wastewater discharges is primarily through the sluice water conveying bottom ash from the boiler. Sluice water has been tested during co-firing of RDF and coal, and results have been compared with data from firing coal only (Gorman, et al., 1977). The reported results for a high ash content RDF indicated that co-firing increased the ash pond effluent concentrations of BOD and suspended solids, reduced dissolved oxygen concentration, and had virtually no effect on oil and grease concentration. The report concludes that additional treatment steps (aeration, flocculation, etc.) may need to be added to the ash pond. Additional treatment may not be necessary in all cases of firing RDF, since the quality of the sluice water supply, the ash content of the RDF, the extent of RDF burnout, and the effluent guidelines differ by location.

Some concern has been expressed about emission of HCℓ that might occur when waste is burned, resulting from certain plastics, aerosols, and food wastes in MSW. HCℓ is corrosive and an irritant, even in small concentrations. If HCℓ emissions become a problem, control will be necessary. Since HCℓ is highly soluble in water it can probably be removed by alkaline water scrubbers. Also, some of the HCℓ reacts with particulate matter and can be removed by dry particulate control systems. However, a study of MSW incineration with and without the addition of large amounts of plastic PVC showed HCℓ emissions were not excessive (Kaiser and Carotti, 1973).

B. Control Regulations

In the absence of other requirements prior to 1960, the guidelines used by combustion engineers and plant operators were those established by the American Society of Mechanical Engineers (ASME). These limited particulate emission to 0.85 lbs per 100 lbs of gas at 50% excess air. Various state and local codes were passed during the early 1960s. In 1966, the Secretary of Health, Education and Welfare was charged with air pollution prevention, control, and abatement from federal government activities. These duties were transferred to the U.S. Environmental Protection Agency (EPA) (1976) and expanded to the country as a whole in the Clean Air Act of 1970. The administrator of EPA was given the task of promulgating national air quality standards. A series of standards followed during 1971, the final two differing primarily in what was to be measured (Gilardi and Schiff, 1972).[1] A marked decrease in allowable particulate emissions took place in approximately one decade. This can be seen in Table VI. Table VII shows the associated conversion factors.

The tightening of standards imposed new equipment requirements. The 1960 ASME requirements generally could be met with baffles and sprays, with good plant operating practices. If particulate matter issuing from a refuse-burning furnace averaged 11 to 22 lbs per ton of refuse charged, an air cleaning efficiency of 11 to 56% would meet the 1960 standard. However, the required efficiency for the 1971 standards is 86 to 93%, requiring more expensive equipment such as electrostatic precipitators, baghouse filters, or scrubbers (Abert, 1977).

C. *Air Emissions Standards for Combined Firing*

There are standards for conventional incinerators, i.e., those which fire 100% unprocessed solid waste with only occasional use of auxiliary fuel for startup or to aid combustion when the fuel is wet, which are separate from standards for conventional fossil fuel boilers. Generally, modification of an old boiler meant that it would then fall under new source standards. There was a question as to whether this would be the case when old boilers—defined as those constructed or modified prior to August 17, 1971—were modified to burn a combination of MSW and conventional fuels.

In 1974, EPA ruled that old boilers modified to burn MSW in combination with fossil fuel need not comply with new source standards. They would, however, remain regulated by the State Implementation Plans (SIP) and be subject to standards promulgated by local authorities. This exemption continued until January 1979, when it became the responsibility of the individual states to decide, based on their EPA-approved SIPs, whether modifications of an old boiler to fire RDF will result in reclassification as a new source.

---

[1]*The 0.08 gr/scf standard does not include those materials which are in a gaseous state at the temperature at which they would be discharged, but would later condense to liquid or solid when cooled to normal temperatures. Presumably, these generally amount to a small part of the allowable particulate emission and account for the difference between 0.10 gr/scf when controlled, and 0.08 gr/scf when not. See Greenberg and Gordon, et al. (1978) and Greenberg and Zoller, et al. (1978).*

TABLE VI. A Decade of Tightening Emission Standards

	lbs/100 lbs flue gas at 50% excess air	gr/scf at 50% excess air	gr/scf at 12% CO_2	lbs/ton of refuse charged
1960 ASME	0.850^a	0.442	0.497	9.58
1966 Federal HEW	0.342	0.178	0.200^a	3.77
1971 Federal EPA	0.362	0.188	0.212	4.00^a
1971 Federal EPA	0.171	0.089	0.100^a	1.88
1971 Federal EPA	0.136	0.071	0.080^a	1.50

[a] Standard given in code.
Source: Adapted from Greenleaf/Telesca (1972).

TABLE VII. Conversion Factors

	lb/ton refuse	lbs/100 lbs flue gas at 50% excess air	gr/scf at 50% excess air	gr/scf at 12% CO_2
lb/ton refuse	1.0	0.09	0.047	0.053
lb/100 lbs flue gas at 50% excess air	11.3	1.0	0.52	0.59
gr/scf at 50% excess air	21.3	1.93	1.00	1.12
gr/scf at 12% CO_2	18.9	1.71	0.89	1.00

[a] Source: W. R. Niessen and A. F. Sarofim (1968).

The particulate emission standard for a new incinerator is 0.08 gr/scf (corrected to 12% CO_2 and 50% excess air).[1] The standard for new fossil-fired steam generating boilers is stricter, 0.10 lb/10^6 Btu (approximately 0.06 gr/scf, corr.).[2] EPA has recently proposed more stringent standards for new electric utility steam generating boilers of 0.03 lb/10^6 Btu (approximately 0.02 gr/scf).[3] This applies to boilers capable of firing in excess of 250 × 10^6 Btu/h (∿10 to 12 tph of coal) of fossil fuels. Below that size, the boiler comes under the SIP for emission standards. Small boilers, however, are still required to obtain Prevention of Significant Deterioration permits and, on a case-by-case basis, may be required to utilize Best Available Control Technology (BACT). While there may not be a specific new source performance standard for a boiler under 250 × 10^6 Btu/h fossil fuel heat input (approximately 22 MW of electrical energy output), emissions must still be sufficiently controlled so that the National Ambient Air Quality Standard (NAAQS) can be met.

EPA has further proposed that a boiler firing less than 250 × 10^6 Btu/h of fossil fuel utilizing solid waste or RDF as a fuel source will be classified as an incinerator for air emission control purposes (the boiler must charge at least 50 tpd of solid waste to come under federal guidelines). In this context, the boiler can be as large as technically feasible using RDF as the primary fuel, without becoming subject to the more stringent air emission standards. Conversely, co-firing RDF with fossil fuel in excess of 250 × 10^6 Btu/h results in reclassification as a process steam generating or electric utility steam generating boiler and subjects the boiler to the more stringent standards applicable to those respective types. Thus a new large electric generating facility must meet all the requisite air quality standards for a utility, whether it co-fires RDF as a supplemental fuel or not.

Under the Clean Air Act Amendments of 1977, EPA promulgated an emission offset policy. A proposed new facility, to be located in an area which does not achieve NAAQS (a nonattainment area), must reduce emissions from other sources already located in that area by an amount greater than the expected emissions from the proposed source. The net result

---

[1] *40 CFR 60.52 (1976), subsection E.*
[2] *40 CFR 60.42 (1976), subsection D.*
[3] *Conversion from lbs/10^6 Btu to gr/scf is approximated by dividing by 1.7. This assumes a pulverized coal-fired boiler burning 12,500 Btu/lb coal with 50% excess air.*

must be an overall improvement in air quality. It is expected, however, that EPA will propose regulations exempting boilers firing 100% RDF. Thus, a new waterwall incinerator would not be required to find offsetting emissions. Being in a non-attainment area, the facility would still be subject to the Lowest Achievable Emission Rate (LAER) requirement, however.

LAER is defined as the most stringent emission limitation for that class or category of source contained in any SIP, unless the source can show that such limitations are not achievable; or the most stringent emission limitation achieved in practice or which can reasonably be expected to occur in practice by that class or category of source. The more stringent of these two options must be applied (Ember, 1978). It is expected that this requirement will result in lower emissions for an incinerator or RDF boiler than the 0.08 gr/scf mentioned above.

D. Mass Burning

It is not necessary to devote a great deal of space to the controlling of emissions in mass burning. There is extensive literature on this subject (e.g., Weinstein and Toro, 1976). In a number of installations this is successfully accomplished, principally by use of electrostatic precipitators. In some cases the initial attempt was not successful, and the retrofit of more effective equipment was necessary.

E. Suspension Firing with RDF with Coal

Most knowledge concerning air emissions and their control from suspension burning results from studies of the St. Louis demonstration project. The air pollution control equipment used there was originally designed and installed for coal only. The demonstration project tested whether it was suitable for combined firing without modification.

One objective of RDF preparation is *quantity* enhancement; i.e., the maximization of the quantity of combustible material in the incoming MSW which after processing becomes the fuel fraction. High velocity air classification can increase the amount of the combustible large particle size material (considering all three dimensions) such as food wastes, wood, leather and rubber, etc., that will be included in the light fraction as fuel. A second shredding step can reduce the particle size and suitability of these materials for suspension burning. Shredding at any stage increases the proportion of the glass, rock, etc., which is shattered into particles sufficiently small to be aerodynamically entrained in the light fraction. Much of this material could be removed prior to shredding or air classification by screening (Alter, 1979a; Woodruff, 1976). Therefore, the finer shredding and higher velocity air classification—while increasing the proportion of the calorific value of the MSW converted to RDF— increases the probability of noncombustible material (ash) also being included.

The alternative is to design the RDF preparation process so as to enhance the *quality* of the RDF. Doing so imposes a penalty. A smaller fraction of the mass of the input waste is delivered as fuel; hence, a higher proportion, and consequently some combustible material ends up as a residue for landfill. If the production of RDF is seen as a means of alleviating a disposal problem, quality enhancement will lack appeal to public works officials. Conversely, it is the goal of the fuel purchaser.

Quality enhancement reduces the content of particulate ash material, or what Fan (1975) called "removable inorganic fines." It would also reduce the quantity of large or heavy pieces which are not likely to burn in suspension. The reduction of ash will increase the calorific value of the fuel as delivered (Sheng and Alter, 1975). The total amount of ash will be reduced, so the boiler operator will have less difficulty in managing the bottom ash. Still at issue is the proportion of the fuel which will burn during the brief suspension time and the proportion of the total ash which reports as fly ash. The latter is particularly important in estimating whether existing air pollution control equipment will be sufficient for combined RDF and coal firing.

Results of the St. Louis demonstration project showed that at the rated capacity of the particular boiler used, particulate emissions measured prior to the electrostatic precipitator were essentially the same whether coal was burned alone or in combination with RDF, even though the ash content of the RDF used was 19% compared to 7.8% for the coal, as shown in Table VIII (Holloway, 1976, Kilgroe, et al., 1975; and Shannon and Gorman, 1975). The RDF used in St. Louis was produced with a view toward maximizing the yield of RDF from the input MSW. The overall bulk density and the size distribution of the St. Louis RDF is shown in Table IX. Unfortunately, no measure of the bulk density distribution of the

TABLE VIII. St. Louis Coal and RDF Characteristics (as received)

	Per lb of fuel		Per 10^6 Btu	
	Coal	RDF	Coal lbs	RDF lbs
Heat value (Btu/lb)	11,000	4,900		
Moisture	13.0%	27.0%	12	55
Ash	7.8%	19.0%	7.1	39
Sulfur	1.2%	0.14%	1.1	0.29
Chlorides	0.38%	0.33%	0.34	0.67

Source: Holloway (1976).

TABLE IX. Average Characteristics of Refuse-Derived Fuel (St. Louis Experience)

Shred Size	wt % passing[a]
<2-1/2 in.	97.5
<1-1/2 in.	94.0
<3/4 in.	73.5
<3/8 in.	49.0

Heating value = 4,900 Btu/lb.
Bulk density = 5.93 lb/ft^3

[a] 2.5 wt % >2-1/2 inch.
Source: Gorman, et al., (1977).

individual particles is available.[1] However, there must have been a significant proportion of "lights" with bulk densities (or other characteristics) which did not permit their combustion during the time they were in suspension. Bottom ash with the combined firing averaged seven times higher than when coal was fired alone. Some of this bottom ash was unburned or partially burned fuel; only 87% of the potential energy in the RDF was released as heat in the boiler (Gorman, et al., 1977; Fiscus, et al., 1976).

The proportion of total ash reporting as fly ash in the St. Louis demonstration was estimated to be 16% (recognizing there is no accepted design method for calculating particulate fly ash from combined suspension firing of refuse and conventional fuels) (U.S. EPA, 1976). The estimates from actual data ranged from 13 to 35%. However, these estimates do not account for the observation that measurements of particulate loading in the gas stream prior to the electrostatic precipitator showed essentially the same values when coal was fired alone. The mid-point of the measurements was the same, 3.8 lbs/10^6 Btu, provided the boiler was not operated above normal capacity. The range was from 3.2 to 4.4 lbs/10^6 Btu (Gorman, et al., 1977; Fiscus, et al., 1976). At the midpoint, the amount of fly ash emitted by the RDF substituted for coal must be essentially the same absolute amount as that of the coal it replaced, or a lower proportion of the total ash of RDF reports as fly ash than for coal. If this holds, then the amount of fly ash produced during RDF/coal co-firing will be reduced (for a given heat input) if the quality of the RDF is enhanced by low velocity air classification and/or screening.

Because of the degree of air cleaning necessary to meet air pollution standards, it is expected that boilers with the capacity to burn significant amounts of RDF will have to be equipped with electrostatic precipitators (ESP). ESP collection efficiency is a function of (1) fly ash particle size,

---

[1] *Using a bulk density as the sole criterion here is probably wrong; the issue involves shape (drag coefficient) and particle sizes in three dimensions. For example, a sheet of paper and a cube of wood, of equal maximum dimension, will be retained on the same size screen. Indeed, a large number of pieces of both may have the same bulk density, or nearly so. However, the paper and wood will burn quite differently in suspension. Even at the same mass burning rate, the paper will be consumed and the wood not. The measures of suitability of the fuel particle must include physical characteristics such as density, maximum size, and aspect ratio.*

(2) ESP power input, (3) the individual particle's ability to accept an electrostatic charge, and (4) the exhaust gas flow rate. Data from the St. Louis project show that when RDF was fired with coal, each of these parameters except ash size changed to decrease the effectiveness of the installed ESP system. However, at rated boiler capacity, the ESP outlet particulate emissions were approximately 0.15 lbs/10^6 Btu for combined firing, which was the same as for coal fired alone. This is an ESP efficiency of approximately 96%.

The EPA requirement for "new source" coal or oil burning steam producing plants is a maximum particulate emission of 0.10 lbs/10^6 Btu fired. If it had to meet the new source standard, the St. Louis plant would have needed air cleaning equipment capable of 97 to 98% efficiency. As a result of the EPA ruling, the St. Louis plant and others making similar conversions were not reclassified as new sources when modified to burn RDF.

F. *RDF, d-RDF, and Stoker Firing with Coal*

An alternative to burning of RDF in suspension is the burning of the material principally on grates as RDF fluff or densified to cubettes or pellets.

An experimental project in Columbus, Ohio, used RDF fluff as a supplement to coal in a stoker boiler. Some data are available on particulate and other discharges from the combined firing in Columbus (Battelle Columbus Laboratories, 1976). One point of this project was to examine the effect of firing low-sulfur refuse with various sulfur content coals to dilute and reduce overall SO_x emissions, and was accomplished to some extent. Combined firing examined RDF/coal mixtures containing as much as 45% RDF by weight (30% by calorific value). Unfortunately, the base line data for coal particulate emissions was not comparable to the data for combined firing. The boiler was fitted with mechanical cyclones as the emission control equipment, was not in compliance, and few conclusions could be drawn.

Air emission measurements were made during trials in England burning mixtures of coal and d-RDF (Birch and Jackson, 1979). The analysis of the fuel used is shown in Table X; the results of the emission measurements for 100% coal, 100% d-RDF, and a 50:50 mixture are shown in Table XI. No mention was made of the type of air pollution equipment fitted on the boiler. The results are reported in somewhat different format than in the U.S. (e.g., SO_4 rather than SO_x). However, sulfate emissions were reduced and chlorides increased by blending the coal with d-RDF. Particulates were lower for 100% of either fuel than for the 50:50 mixture.

TABLE X. Analysis of Fuel Used in Stoker Tests

	Pellets	Coal (Arkwright washed smalls)
Proximate analysis (wgt %)		
Moisture	12.5	6.1
Ash	10.2	6.5
Volatile matter	65.4	34.0
Fixed carbon	11.9	53.4
Volatile matter (dry, ash-free % basis)	84.6	38.9
Ultimate analysis (wgt %)		
Moisture	12.5	6.1
Ash	11.7	6.9
Carbon	49.8	76.8
Hydrogen	6.1	5.1
Sulfur	0.2	1.65
Chlorine	0.24	0.22
Carbon dioxide	0.22	0.08
Calorific value		
Btu/lb (dry, ash-free basis)	9,370	14,790

Source: Birch and Jackson (1979).

TABLE XI. Concentrations of Pollutants from Stoker Tests

Fuel	$SO_4^=$ mg/m^3	Cl^- mg/m^3	NO ppm	NO_2 ppm	Total hydrocarbons ppm C_3	Particulates mg/m^3
100% coal	2366	12.1		80.7[a]	14	31.0
100% RDF pellets	188	25.3	114	6	11	
100% RDF pellets	200	104.7	110	5	11	70.1
50% RDF/50% coal	1765	54.5	132	2	15	
50% RDF/50% coal	1757	40.6	140	1	15	82.9

[a] $NO + NO_2$

Source: Birch and Jackson (1979).

VII. CORROSION

Predicting the extent of corrosion resulting from the firing of unprocessed waste or from some form of RDF, singularly or in combination with another fuel, is extremely difficult. Metal wastage occurs on waterwalls near the flame zone, in lower temperature gas passes, in air heaters and in the high temperature superheater and reheater sections of various types of boilers.

The mechanisms of corrosion believed to occur are the same as from fossil fuels. Liquid phase corrosion from molten ferric chloride or sodium oxide occurs on boiler tube surfaces. Dew point or acid corrosion takes place in the low temperature gas passes from the condensation of sulfuric or sulfurous acid. Means of reducing corrosion have included greater use of refractoring shields even on boiler tubes, new alloys, and careful attention to operating procedures. For a discussion of recent findings, see Matula (1977).

Corrosive wastage should not be confused with erosion. The latter is caused by high velocity, particulate-bearing gas streams passing through tube banks. If RDF is burned in a boiler originally designed for coal, or worse oil, the tube spacing would have been designed for lower total gas volumes (hence lower velocities) than will be experienced from RDF. Erosion can be reduced or avoided by including tube shielding, increasing tube shielding, increasing tube spacing, changing the location of soot blowers, and the like (Velzey, 1979). Operating and design experiences for several installations have been reported (for example: Stabenow, 1976; Velzey, 1979). European experience is available to operators in the U.S., generally through license of a particular incinerator design.

Refuse is a heterogeneous, nonspecification fuel. One objective of processing it to some type of RDF is to reduce its heterogeneity of both form and composition, and thus reduce its propensity to corrode. The corrosion to be expected from unprocessed refuse, burned in an incinerator of some type, is likely to be different and more severe than from burning RDF. There are several field reports supporting this difference but little, if any, documentation.

Part of the reason for the high corrosion observed in mass burning incinerators is that refuse is heterogeneous, both in physical form and composition. The physical form makes it difficult to adequately distribute the combustion air, particularly underfire air. If this occurs, the combustion gases may be rich in carbon monoxide which is a reducing atmosphere for hot steel surfaces. Subsequent shifts to excess air, and oxygen in the combustion gases, can oxidize the same metal surfaces resulting in high metal wastage.

The composition heterogeneity also contributes to corrosion. For example, one result for the ultimate analysis of refuse was: sulfur 0.26, chlorine 0.77, phosphorous 0.10, wt % dry basis (Alter, et al., 1974). In contrast, a bituminous coal might contain sulfur ≃1, chlorine 0.05 or more, and phosphorous 0.02 wt %. Potassium and sodium compounds in the refuse can contribute to corrosion and the formation of slag deposits on heat exchange surfaces. These deposits may be corrosive in themselves and/or may entrap other corrosive compounds, such as ferric chloride.

Chlorine content is important. Also of concern is whether the chlorine is inorganically bound. If the chlorine is in the form of sodium chloride (NaCl), is not likely to decompose at boiler temperatures, although it may react chemically. Organically bound chloride is labile and likely to decompose to corrosvie compounds. The familar plastic polyvinyl chloride (PVC) decomposes far below boiler temperatures and liberates about half its weight as hydrochloric acid (HCl). This acid can react in complex ways with metal surfaces to cause corrosion. Fortunately, waste contains little PVC and processing to some form of RDF reduces the PVC content of the fuel even further.

There is speculation that one of the chemical reactions accounting for boiler corrosion is as follows:

$$2\ NaCl + SO_2 + 1/2\ O_2 + H_2O \quad Na_2\ SO_4 + 2\ HCl$$

The Na_2SO_4 can be accounted for in slag deposits; the HCl is the corrosive agent (Kerekes, et al., 1977). There is no known direct evidence that this reaction occurs. It has been shown to be thermodynamically possible (Alter, 1979b). It has been proposed that if the ratio of sulfur to chlorine can be controlled, however, the HCl will be maintained in the gas phase and not cause corrosion. When a form of RDF was burned with various coals of differing sulfur contents, a minimum corrosion was found for the predicted ratio of S/Cl (Vaughan, et al., 1976).

The NaCl content of both unprocessed refuse and RDF is uncontrollable except possibly by attention to the S/Cl ratio. However, the chlorine content of a properly prepared RDF (i.e. the quality enhancement referred to earlier) is expected to be lower than for other forms of RDF or unprocessed MSW. One reason is the distribution of chlorine (and some other potential acid forming elements) in various constituents of refuse. For example, in one sample of refuse, 33% of the chlorine was found in constituents expected to be included in an air classifier heavy fraction, not counting the plastics. The plastics in the waste fraction accounted for 36% of the total chlorine;

some of these, such as moulded or extruded PVC items, would also be expected to be included in the heavy fraction (Alter, et al., 1974). Klumb (1976) reported that approximately 80% of the chorine in the RDF prepared in St. Louis was water soluble and accounted for as NaCℓ.

In St. Louis, the RDF burned had a higher total or organic chlorine content than the coal used. Further, the sulfur content of the RDF was 10% of the sulfur in coal (Table VIII). Co-firing RDF and coal can dilute the total sulfur and chlorine content as illustrated by Tables I and X.

A key factor in controlling corrosion is limiting the steam temperature and pressure. Systems burning unprocessed MSW generally have been limited to relatively low temperatures and pressures. However, the plant in Saugus, Mass., and some European installations do superheat the steam, a practice initially beset by difficulties. Representative operating conditions are steam conditions of about 600 psig/1560°F. Tube metal temperatures may be 900°F. These conditions can lead to high corrosion with some types of coal or refuse fuel.

Some studies found that while the corrosion rate of low alloy steels increased linearally with the amount of chlorine present in the waste at temperatures above 900°F, the corrosion rate of stainless steels were not so affected, but increased with gas temperatures (Vaughan, et al., 1976).

There is a trend toward higher steam temperatures and pressures in units burning RDF. This can be seen from Table XII. For example, the plant in Niagara Falls is designed for 1200 psig and 750°F, burning 100% RDF. Utility plants co-firing coal and RDF operate at around 1500 psig or higher, and 1000°F.

VIII. ECONOMICS

The economics of resource recovery systems are based on three factors: (1) capital and operating costs of the technique employed, (2) revenue from the recovered products, and (3) tipping fees. Tipping fees might be derived from the general budget of the community, but if so, can be converted into a per-ton disposal charges. Tipping fees vary among communities and are usually governed by the cost of alternative methods of disposal. Some communities may be willing to pay more for resource recovery than for alternative methods for environmental or conservation reasons. More likely, if there is a willingness to pay more, it will be to avoid potential difficulties of siting a new landfill or concluding a regional agreement to ship the waste out of the affected community. In general, the resource recovery tipping fee has

TABLE XII. Pressure, Temperature, and Degrees of Superheat Boilers Using Refuse-Derived Fuel

	Pressure (psig)	Temperature (°F)	Degrees superheat (°F)
St. Louis	1250	950	376
Akron[a]	560	479	0
Hamilton	250	405	0
Niagara Falls[a]	1200	750	182
Columbus[a]	700	725	220
Hempstead[a]	600	750	262
Dade County[a]	600	750	262
Milwaukee	2620	1050	375
Chicago	2000	1000	364
Monroe County[a]	2000	1050	414
	1475[b]	955[b]	359[b]
	1825[b]	1000[b]	375[b]
Bridgeport[a]	2000	1000	764
	1500	1000	403
Newark[a]	2350	1100	441
Hagerstown	150	366	0

[a] Planned or under construction. Others on-line or in shakedown.
[b] Four boilers— range between.

an upper boundary approximating the cost of the least expensive alternative method of disposal, usually landfill.

For most approaches to resource recovery, energy is the principal revenue product. In general, the more that is invested in the system, the higher the revenue from the energy product. Thus RDF has a lower market value than steam, and steam a lower market value than electricity.

Recently, resource recovery advocates have begun to point out that with the expected increase in energy value and the expected increase in capital cost of new plant and equipment, an investment made now in energy recovery plant and equipment appears to have an interesting financial future. Figure 6 illustrates this. The key to interpreting the figure is the horizontal line depicting the amortized capital cost. The fact that it is level is the key to the positive picture for recovery depicted by the figure. An investment, once made, is paid back in level installments. However, other costs and revenues increase. Operating costs are shown increasing at 6% per year. Revenues from energy sales are depicted increasing at 9% per year. The tipping fee, or alternate disposal cost, is shown increasing at 6% per year.

On this basis, the breakeven point is about five years; net plant costs for that particular year equal alternative disposal costs. From then on, recovery is a less expensive option than alternative disposal approaches. At time zero, the graph assumes this to be $10 per ton. This is realistic for some of the major metropolitan areas. The graph has to be adjusted to take into account a lower cost in other areas. This does not detract from the point of the graph portraying economic "readiness" for resource recovery. Moreover, the program costs, in a life cycle sense, would indicate that a community with the cost structure depicted by the figure would save money over a 20 year program. Such a community would be ahead overall by about the eighth year. This is when the accumulated losses, measured against the cost of alternative disposal costs, would be offset by the accumulated savings measured against the same (but escalating) standard.

The energy sales line in Figure 6 depicts a generalized energy value. The most significant determinant of the value of refuse-based energy is the alternative fuel that would have been used if refuse were not available. The more expensive the alternative fuel, the higher the value that can be assumed for refuse when used as a substitute. Under present prices of conventional fuels, it is better to substitute for oil than for coal. If oil prices continue to escalate and coal does not follow at the same, or roughly the same rate, then the economics of offering solid waste as an oil substitute will become even more attractive.

The two most likely forms of waste energy or resource recovery products are a RDF or steam. The former may be burned with a conventional fuel, most likely coal, as a substitute for part of the normal fuel used. The latter may be generated by burning the RDF separately, as 100% of the fuel in a dedicated boiler. Not surprisingly, the market price (in terms of $/energy unit) is higher in the steam case, but the costs of the production of the steam are higher than the cost of the projection of just fuel. The significant advantage of steam is in marketing. Because the refuse processo

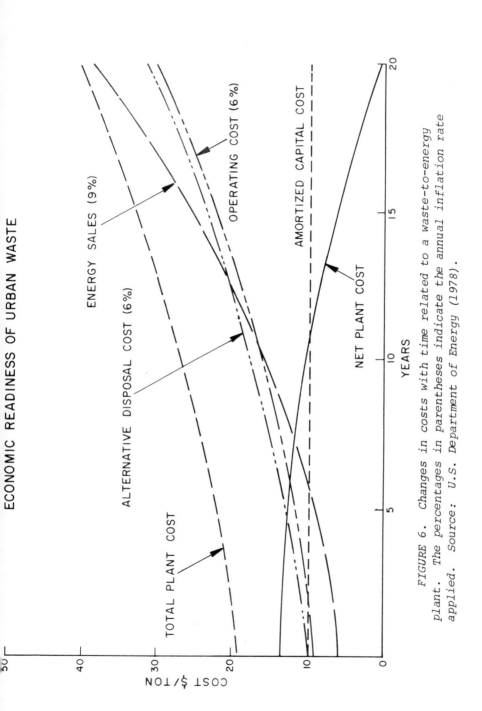

FIGURE 6. Changes in costs with time related to a waste-to-energy plant. The percentages in parentheses indicate the annual inflation rate applied. Source: U.S. Department of Energy (1978).

takes the risk of producing the steam, it can generally obtain a firm long-term commitment to sell the steam to an appropriate user—if one is located in his area. In the case of fuel production and sale, the risk mainly falls on the purchaser. In the current time frame, because of the lack of operating experience with RDF, users are reluctant to sign firm long-term contracts, especially on a take-or-pay basis.

Table XIII shows ranges of likely revenues for (1) the sale of refuse-derived fuel, and (2) the sale of steam. The basis for the estimation is the market price of the competitive fuel. For convenience, no market incentive (e.g., a price discount) for the purchaser is shown. In actual practice, a refuse-derived energy purchaser will expect some economic advantage from his involvement or else it would not be logical to abandon tried, tested, and true fuel or steam sources. It is easy enough to adapt the information in the table to the idea of a discount.

TABLE XIII. Potential Energy Revenues

Refuse derived fuel	- 12×10^6 Btu/ton - 60% of input municipal waste		
Market price of competing fuel ($/$10^6$ Btu)	$0.50	$ 1.00	$ 1.50
Revenue per input ton of municipal waste	3.60	7.20	10.80
Steam	- by incineration of raw waste (10×10^6 Btu/ton) - 50% efficiency of incinerator/boiler; 5000 lbs steam per input ton waste		
Market price of steam ($/thousand lbs)	$1.00	$ 2.00	$ 3.00
Revenue per input ton of municipal waste	5.00	10.00	15.00

IX. RESEARCH NEEDS

As an increasing number of refuse-burning resource recovery facilities are built, tested, and begin to operate on a regular basis, the needs for laboratory—or full scale—research in a number of important areas has become more sharply defined. In the last several years seminars, workshops, and publications have dealt with the critical questions of what type of basic knowledge, experimental and operational data are needed to further resource recovery technology.

In 1976, a conference resulted in the publication of a monograph on the "Present Status and Research Needs in Energy Recovery from Waste" (Matula, 1977). Another report examines research needs in "preprocessing," understood to be the set of operations preceding the production of the fuel fraction itself (Bendersky, et al., 1976). There are others, as well (Franklin, 1977; Smithson, 1978). An EPRI/EEI workshop in 1977 had a section on "Technical Issues" which affect the use of RDF by utilities, with a list of subjects calling for more research on a priority basis (Electric Power Research Institute, 1977). Table XIV presents a list of work needed to accelerate energy from waste. The paragraphs below aggregate the suggestions resulting from the conferences and publications cited.

"Preprocessing" of waste to produce RDF or d-RDF. The main areas of need appear to be: better knowledge of characteristics, influence of screening, shredding, and air classification on combustible fraction quality and quantity; effects of moisture content and fuel drying; conveying, storage, and handling methods and facilities for RDF and d-RDF; methods for densifying RDF.[1]

Fuel properties of RDF and d-RDF. More information is needed on methods for sampling, analyzing and classifying waste fuels. There is need for further development, refinement, adaptation, or invention.[2]

Effluent streams measurement, characterization, and control. More information is needed on the characteristics

---

[1] *"Preprocessing" is a non sequitur and a better term is needed.*

[2] *This is one of several areas of investigation being conducted by Committee E-38, Resource Recovery, of the American Society for Testing and Materials, under contract to the U.S. Environmental Protection Agency. See Hollander and Kiefer (1978) and Alter (1979c).*

of effluents from the furnace, into the air pollution control equipment, and into the stack.

Pollution control equipment. Areas of specific interest include: performance and applicability of wet scrubbers and bag filters, full-scale tests of equipment, development of new materials.

Steam generation processes and equipment. The main need for R & D concern:

(1) Corrosive deposits on high-temperature superheater and reheater surfaces, more specifically: high metal temperature limits, the role of heavy metals, the short- and long-term effects of chlorine, reactions on tube surfaces, long-life alloys.

(2) Waterwall deposits and corrosion due to lower melting point slags from RDF and chlorine release near the tube surfaces.

(3) Low temperature corrosion due to acids and stress corrosion fatigue.

(4) Combustion: flame length suppression, devices and methods for the prediction of RDF combustion characteristics, operating data for reliable mixes of known RDF and other fossil fuels in various systems, load-swing and turn-down capability.

(5) Slagging and fouling: rate vs. fuel specification and mixture ratio, effects of sootblowing.

(6) Ash: quantities vs. fuel specifications for different firing arrangements.

The above list of topics and Table XIV are certainly not exhaustive, but simply indicative of present trends and thoughts of researchers and systems operators in the use of refuse as an energy source.

ACKNOWLEDGMENT

Portions of this chapter have been taken from previous work of the authors and of other employees of the National Center, in particular: "Refuse-Derived Fuel...Energy for Industry," NCRR *Bulletin,* Vol. V, No. 2 (Spring 1975), pp. 39-57, authored by J. G. Abert; "Air Pollution from Burning Refuse Fuels," NCRR *Bulletin,* Vol. VII, No. 1 (Winter 1977), pp. 15-28, by J. G. Abert; and "Refuse-Derived Fuel (RDF and Densified Refuse-Derived (d-RDF)," NCRR RM 77-2 (June 1978), 67 pages, by Marc L. Renard. The table in the Appendix of this chapter is a regular publication of NCRR and is prepared by Christina Giambi.

TABLE XIV. Research Needs

TECHNOLOGIES

To improve the technology of mass burning waterwall furnace incineration with steam of electricity generation.

To advance the state of the art of dry shredded (and set pulped) fuel preparation systems.

To further the development of pyrolysis, including oil and syngas generation as well as methanol/ammonia recovery, as an efficient, economic, and environmentally sound energy recovery technology.

To develop fluidized bed combustion technology into an economical, efficient, environmentally sound method for energy recovery from various wastes; and to conduct research on the application of gas turbine systems.

To advance the state of the art of anaerobic digestion of wastes, and to increase its energy recovery capability.

To develop modular and semi-portable energy recovery systems for small-scale applications.

To improve the technologies for recovery of materials from the waste stream.

PRE-PROCESSING

To quantify municipal, agricultural, industrial, and forestry wastes available for processing by type and geographic distribution.

To identify those constituents in the waste streams which are particularly troublesome to processing and processing equipment.

To investigate innovative alternative approaches to conventional methods of waste particle size reduction relative to intended use, energy consumption and overall economics.

To determine the preferred waste feedstock particle size for selected energy recovery technologies.

To investigate equipment separation efficiencies and energy trade-offs.

To improve the processability of waste materials in terms of combustion characteristics and decomposition characteristics.

To investigate the feasibility of simultaneous processing of different waste types.

(Continued)

TABLE XIV. (Continued)

UTILIZATION

To expand support of systematic analysis of technical/economic issues surrounding the use of waste-based products in terms of the development of standardized product specifications and the development of standard analytical procedures to determine physical and chemical characteristics of recovered products.

To assess the marketability potential of waste-based fuels.

To improve storability and transportability of waste-based fuels.

To perform economic analyses of retrofitting various types of boilers with grates, ash handling capability, nozzles, and other equipment necessary to permit the use of waste-based fuels.

To perform feasibility studies on cocombustion of oil and shredded solid waste, oil and waste-derived oil, and oil with spent oils and lubricants.

To investigate potential effects of waste-based fuels on boiler operations in terms of corrosion, slagging, fouling, and air pollution.

To determine applicability and efficiency of various air pollution control devices to specific processing technologies.

To improve sampling and measurement techniques for emissions from energy recovery processes.

To investigate the quality of liquid effluents resulting from energy recovery processes.

To determine the quality of process residues and potential for reuse.

To investigate potential health hazards which may be associated with processing of wastes.

Source: Adapted from MITRE 1975, "Energy Conservation Waste Utilization Research and Development Plan," MTR-3063, 189 pp.

REFERENCES

Abert, J. G. (1975). National Center for Resource Recovery *Bulletin*, V (2), pp. 39-57.

Abert, J. G. (1977). National Center for Resource Recovery *Bulletin*, VII (2), pp. 15-28.

Alter, H., Ingle, G., and Kaiser, E. R. (1974). *Solid Waste Management* (England), 64 (12, pp. 706-712.

Alter H. (1977). *Environmental Conservation 4*, pp. 11-20.

Alter, H. (1977a). *In* "Present Status and Research Needs in Energy Recovery from Wastes" (R. A. Matula, ed.), pp. 9-22. American Society of Mechanical Engineers, New York.

Alter, H., and Dunn, J. J. (1979). "Solid Waste Conversion to Energy: Current European and U. S. Practice," M. Dekker, New York.

Alter, H., ed. (1979a). "Materials Recovery from Municipal Solid Waste: Investigation of Air Classification, Upgrading RDF, Aluminum and Glass Recovery," U.S. Environmental Protection Agency (Final Report: Grant 803901), Cincinnati, Ohio.

Alter, H. (1979b). *In* "Proc. Symp. on Materials from the Chemical Viewpoint" (E. Horowitz and L. Shubert, eds.). American Chemical Society, Washington, D.C.

Alter H. (1979c). *Conservation and Recycling, 2*.

Appell, H. R., Fu, Y. C., Friedman, S., Yavorsky, P. M., and Wender, I. (1971). "Converting Organic Wastes to Oil," U.S. Bureau of Mines, Report of Investigations 7560.

Autem, W. H. (1975). Private communication.

Babcock and Wilcox Co. (1963). "Steam: Its Generation and Use," Babcock and Wilcox, New York, New York.

Babcock and Wilcox (1973). "Hydropulped Municipal Waste Burning Test for Black Clawson Company." Private communication.

Battelle Columbus Laboratories (1976). "Environmental Effects of Utilizing Solid Waste as a Supplementary Power Plant Fuel," Battelle Columbus Laboratories, Columbus, Ohio.

Bendersky, D., Keyes, D. R., Luttrell, M. G., Simister, M., and Viscek, D. (1976). "Study of Preprocessing Equipment for Waste-to-Energy Systems, State of the Art and Research Needs," Midwest Res. Inst., Kansas City, MO.

Beningson, R. M., Rogers, K. J., Lamb, T. S., and Nadkarni, R. M. (1975). In Proceedings, 1st International Conference, "Conversion of Refuse to Energy," Montreaux.

Benziger, J. B., Bortz, B. J., Neamatalla, M., Szostak, R. M., Westerhoff, G. R. P., and Schultz, H. W. (1976). "Resource Recovery Technology for Urban Decision Makers," Columbia University, New York.

Birch, P. R. and Jackson, D. V. (1979). In Proceedings 2nd World Recyling Congress, Manila, Philippines.

Bogot, A. and Sherrill, R. C. (1976). *Combustion* 47 (9), 13.

Bracken, B. D., Coe, J. R., and Allen, T. D. (1976). "Full-Scale Testing of Energy Production from Solid Waste," presented at the Third National Conference on Sludge Management, Disposal and Utilization, Miami, Florida.

Buonicore, A. J. and Waltz, J. P. (1975). "District Heating with Refuse-Derived Fuel at Wright-Patterson Air Force Base," Air Force Facilities Command, Dayton, Ohio, unpublished.

Carlson, D., Spencer, D., and Christensen, H. (1976). In Proceedings, 5th Mineral Waste Utilization Symposium, pp. 196-203.

Carter, J. D. (1974). *Paper Trade Journal*, Feb. 13, p. 535.

Chapman, R. A. (1975). In Proceedings, 1st International Conference, "Conversion of Refuse to Energy," Montreux, pp. 343-348.

Chapman, R. A. and Wocasek, F. R. (1974). In Proceedings, 1974 National Incinerator Conference, ASME, pp. 347-357.

Duckett, E. J. (1977). National Center for Resource Recovery *Bulletin VII* (4), pp. 94-100.

Edison Electric Institute (1977). "A Compilation of Case Studies on Solid Waste Utilization Projects Involving Investor Owned Utilities," Edison Electric Institute, New York.

Electric Power Research Institute (1977). "Utilization of Refuse-Derived Fuels by Electric Utilities," Electric Power Research Institute, Palo Alto, Calif.

Ember, L. R. (1978). *Environ. Sci. & Techn. 12,* p. 1008.

Engdahl, R. B. (1976). "Identification of Technical and Operating Problems of Nashville Thermal Transfer Corporation Waste-to-Energy Plant," U.S. Department of Energy, Washington, DC.

Even, J. C., Adams, S. K., Gheresus, P., Joensen, A. W., Hall, J. L., Fiscus, D. E., and Romine, C. A. (1977). "Evaluation of the Ames Solid Waste Recovery System, Part I - Summary of Environmental Emissions: Equipment, Facilities, and Economic Evaluation," U. S. Environmental Protection Agency, Cincinnati, Ohio.

Fan, D. H. (1975). *Resource Recovery and Conservation, 1,* pp. 141-150.

Fenton, R. (1975). *Resource Recovery and Conservation, 1,* pp. 167-176.

Fiscus, D. E., Gorman, P. G., and Kilgroe, J. D. (1976). "Bottom Ash Generation in a Coal-Fired Power Plant when Refuse-Derived Fuel Is Used," Midwest Res. Inst., Kansas City, Missouri.

Flintoff, F. (1978). *Solid Wastes, 68* (9), pp. 349-355.

Franklin, M. A., ed. (1977). "The Preparation of Fuels and Feedstocks from Municipal Solid Waste," Environmental Protection Agency, Cincinnati, Ohio.

Garbe, Y. M. and Levy, S. J. (1976). "Resource Recovery Plant Implementation Guides for Municipal Officials. MARKETS," U.S. Environmental Protection Agency, Washington, DC.

Gilardi, E. F. and Schiff, H. F. (1972). *In* Proceedings, 1972 National Incinerator Conference, ASME, p. 109.

Glaus, Pyle, Schomer, Burns, and DeHaven (1974). "Preliminary Phase Report of Solid Waste Reduction Energy Recovery," Glaus, Pyle, Schomer, Burns and DeHaven, Akron, Ohio.

Godfrey, K. A. (1976). *Civil Engineering, 46* (9), pp. 86-93.

Gorman, P. G., Shannon, L. J., Schrag, M. P., and Fiscus, D. E. (1977). "St. Louis Demonstration Project Final Report: Power Plant Equipment, Facilities and Environmental Evaluations," Midwest Research Institute, Kansas City, Missouri.

Greenberg, R. R., Zoller, W. H., and Gordon, G. E. (1978). *Environ. Sci. & Techn., 12,* pp. 566-573.

Greenberg, R. R., Gordon, G. E., Zoller, W. H., Jacko, R. B., Neuendorf, D. W., and Yost, K. J. (1978). *Environ. Sci. & Tech. 12,* 1329-1332.

Greenleaf and Telesca (1972). "Detail Engineering and Economic Report, Solid Waste Collection and Disposal, Metropolitan Dade County, Florida," *2,* V-38. Greenleaf and Telesca, Miami, Florida.

Hecklinger, R. S. (1976). *In* Proceedings, 1976 National Waste Processing Conference, ASME, pp. 133-140.

Hollander, H. I. and Cunningham, N. F. (1972). *In* Proceedings, 1972 National Incinerator Conference, ASME, pp. 77-86.

Hollander, H. I. and Kiefer, J. K. (1978). Paper presented at 1978 ASME Annual Meeting.

Holloway, R. J. (1976). *Waste Age*, 7 (8), 50-52.

Industry Trends (1977). *Environ. Sci. & Tech.* 11, p. 517.

Jackson, J. W. (1976). "A Bioenvironmental Study of Emissions from Refuse-Derived Fuel," National Technical Information Service, Springfield, Virginia.

Jenn, J. J. and Peters, F. A. (1971). "Cost Evaluation of a Metal and Mineral Recovery Process for Treating Municipal Incinerator Residues," U.S. Bureau of Mines, Washington, D.C., Information Circular 8533.

Jones, J. L. and Radding, S. B. (1978). "Solid Wastes and Residues, Conversion by Advanced Thermal Processes," American Chemical Society, Washington, DC.

Kaiser, E. R. and Carotti, A. A. (1971). "Municipal Incineration of Refuse with 2% and 4% Additions of Four Plastics," Society of the Plastics Industry, New York.

Kerekes, Z. E., Bryers, R. W., and Sommerlad, R. E. (1977). *In* "Present Status and Research Needs in Energy Recovery from Wastes," (R. A. Matula, ed.). American Society of Mechanical Engineers, New York, pp. 120-170.

Kenyon, R. A. (1975). *In* Proceedings, 1st International Conference, "Conversion of Refuse to Energy," Montreux, pp. 256-61.

Kilgroe, J. D., Shannon, L. J., and Gorman, P. G. (1975). *In* Proceedings, 1st International Conference, "Conversion of Refuse to Energy," Montreux, pp. 190-196.

Kispert, R. G., Sadek, S. E., and Wise, D. L. (1976). *Resource Recovery and Conservation*, 1, pp. 245-256.

Klumb, D. (1976). *Resource Recovery and Conservation*, 1, pp. 225-234.

Maryland Environmental Service (1978). "Energy from Wastes," Maryland Environmental Services, Annapolis, Maryland.

Matula, R. A. (ed.) (1977). "Present Status and Research Needs in Energy Recovery from Wastes," American Society of Mechanical Engineers, New York.

Mean, D. (1977). *Solid Waste Management News,* November 1977.

Meissner, H. G. (1969). *In* "Principles and Practices of Incineration," (R. C. Corey, ed.). Wiley-Interscience, New York, pp. 163-209.

Mitre Corporation (1975). "Energy Conservation Waste Utilization, Research and Development Plans," Report MITR 3163, Mitre Corporation, Bedford, Mass.

National Center for Resource Recovery (1975). *Bulletin, V* (1) pp. 2-9.

National Center for Resource Recovery (1979). "Resource Recovery Activities, A Status Report," (C. Giambi, ed.), National Center for Resource Recovery, Washington, DC.

Niessen, W. R. and Sarofim, A. F. (1968). *In* Proceedings, 1968 National Incinerator Conference, ASME, pp. 174-176.

Niessen, W. R. and Chansky, S. H. (1970). *In* Proceedings, 1970 National Incinerator Conference, ASME, pp. 8-16.

Niessen, W. R., Smith, E., and Gilardi, E. (1976). "A Review of Techniques for Incineration of Sewage Sludge with Solid Waste," U.S. Environmental Protection Agency, Cincinnati, Ohio.

Novak, R. G., Cuday, J. J., Denove, M. B., Standifer, R. L., and Wars, W. E. (1977). *Chemical Engineering, 84* (10), p. 131.

Office of Solid Waste Management Programs (1973). "First Report to Congress, Resource Recovery and Source Reduction," U.S. Environmental Protection Agency, Washington, DC.

Office of Solid Waste Management Programs (1974). "Second Report to Congress, Resource Recovery and Source Reduction," U.S. Environmental Protection Agency, Washington, DC.

Office of Solid Waste Management Programs (1975). "Third Report to Congress, Resource Recovery and Waste Reduction," U.S. Environmental Protection Agency, Washington, DC.

Office of Solid Waste Management Programs (1977). "Fourth Report to Congress, Resource Recovery and Waste Reduction," U.S. Environmental Protection Agency, Washington, DC.

Pfeffer, J. T. (1976). *Resource Recovery and Conservation, 1,* pp. 295-314.

Renard, M. L. (1978). "Refuse-Derived Fuel (RDF) and Densified Refuse-Derived Fuel (d-RDF)," National Center for Resource Recovery, Washington, DC.

Renard, R. L. and Fretz, R. (1979). "Study of Burning of Refuse-Derived Fuel in Small Oil Fired Boilers," U.S. Department of Energy, Washington, DC, Final Report, Contract ES-76-01-3851.

Resource Planning Associates, Inc. (1977). "European Waste-to-Energy Systems, An Overview," National Technical Information Service, Springfield, Virginia.

Rigo, G. R., Olexsey, R., and Wiles, C. (1978). "Substituting d-RDF for Coal in an Industrial Spreader Stoker," prepared for presentation at the 6th Mineral Waste Utilization Symposium, unpublished.

Schwieger, R. G. (1975). *Power, 119* (2), pp. S1-24.

Shanks, H. R., Hall, J. L., and Joensen, A. W. (1977). "Environmental Effects of Burning Solid Waste as Fuel," 4th Joint Conference Sensing Environmental Pollutants, New Orleans, Louisiana.

Shannon, L. J. and Gorman, P. G. (1975). "Union Electric Coal-Refuse Firing Boiler: Compilation of Environmental Test Results, October 1974 to July 1975," Midwest Research Institute, Kansas City, Missouri.

Sheng, H. P. and Alter, H. (1975). *Resource Recovery and Conservation, 1,* pp. 85-94.

Sieger, R. B. and Moroney, P. M. (1977). "Incineration-Pyrolysis of Wastewater Treatment Plant Sludges," Brown and Caldwell, Walnut Creek, Calif.

Singer, J. G. and Mullen, J. F. (1974). *Combustion, 45* (8), pp. 20-30.

Smithson, G. R. (1978). "Final Report on Workshop Waste-to-Energy Technology," Battelle Columbus Laboratories, Columbus, Ohio.

Stabenow, G. (1976). *In* Proceedings of 1976 National Waste Processing Conference, ASME, pp. 81-96.

Stern, A. C., Wohlers, H. C., Boubel, R. W., and Lowry, W. O. (1973). Fundamentals of Air Pollution." Academic Press, New York, p. 466.

Systems Technology Corporation (1975). "A Technical, Environmental and Economic Evaluation of the Wet Processing System for the Recovery and Disposal of Municipal Solid Waste," National Technical Information Service, Springfield, Virginia, p. 217.

Unweltbundesamt (1978). "CREST Study of Household Waste Sorting Systems," Final Report, Umeweltbundesamt, Berlin, p. 3.

U.S. Department of Energy (1978). Briefing materials.

U.S. Environmental Protection Agency (1973). "Municipal Scale Incinerator Design and Operation," U.S. Environmental Protection Agency, Washington, DC, p. 57.

U.S. Environmental Protection Agency (1975). "Effluent Guidelines Report, Steam Generating Installations, Development Document," U.S. Environmental Protection Agency, Washington, DC.

U.S. Environmental Protection Agency (1976). "Performance of Emission Control Devices on Boilers Firing Municipal Solid Waste and Oil," U.S. Environmental Protection Agency, Research Triangle Park, N.C., pp. 16-17.

Vaughan, D. A., Krause, H. H., and Boyd, W. K. (1976). In "Present Status and Research Needs in Energy Recovery from Waste," (R. A. Matula, ed.). American Society of Mechanical Engineers, New York, pp. 182-209.

Velzy, C. O. (1979). *Resource Recovery and Conservation 4.*

Weinstein, N. J. and Toro, A. F. (1976). "Thermal Processing of Municipal Solid Waste for Resource and Energy Recovery." Ann Arbor Science, Ann Arbor, Michigan.

Weston, R. F. (1971). "Evaluation of Solid Waste Cubettes as Salvage Fuel for Municipal Electric Generation," Roy F. Weston, Inc., Philadelphia, Pa.

Wise, D. L., Wentworth, R. L., Augenstein, D. C., and Cooney, C. L. (1978). *Resource Recovery and Conservation, 3,* pp. 41-60.

Woodruff, K. L. (1976). *Transactions, Society of Mining Engineers, AIME 260,* pp. 201-204.

APPENDIX

TABLE I.
RESOURCE RECOVERY ACTIVITIES   JANUARY 1979

Location	Key Participants	Process	Output	Reported Capacity	Reported Capital Costs (millions of $)	Status
Akron, Ohio	City of Akron; Glaus, Pyle, Schomer, Burns & De Haven; Ruhlin Construction Co.; Babcock & Wilcox Co. (boiler supplier); Teledyne National (operator)	Shredding; air classification; magnetic separation; burning of refuse-derived fuel (RDF) product in semi-suspension stoker grate boiler	Steam for urban heating and cooling and industrial use; magnetic metals	1000 tons per day (TPD)	46[a]	Under construction; three-fourths complete; in shake-down by July 1979; fully operational by Jan. 1980
Albany, N.Y.	City of Albany and 10 surrounding communities; Smith and Mahoney (designers and project managers)	Shredding; magnetic separation; combustion in semi-suspension stoker grate boiler; recovery of nonferrous from boiler ash	RDF; magnetic metals; steam for urban heating and cooling; nonferrous metals	750 TPD	22	Groundbreaking held in Oct. 1977; construction 60% complete; in operation by Spring 1980
Ames, Iowa	City of Ames; Gibbs, Hill, Durham & Richardson, Inc. (designer)	Baling (waste paper); shredding; magnetic separation; air classification; screening; other mechanical separation	Refuse-derived fuel for use by utility; baled paper; magnetic metals; aluminum, other non-magnetic metals	200 TPD (50 tons per hour [TPH])	6.19[b]	Operational since 1975
Baltimore, Md.*	City of Baltimore; EPA	Landgard® process: shredding, pyrolysis, water quenching, magnetic separation	Steam; magnetic metals; glassy aggregate	1000 TPD	EPA-7 State of Maryland - 4 City of Baltimore - 11 Monsanto - 4 Additional funds: Dept. of Commerce, F.E.D.A. - 3.1 City of Baltimore - 1	Monsanto Enviro-Chem Systems, Inc., has withdrawn from the project; plant temporarily closed for installation of air pollution control equipment and other modifications; startup scheduled in early 1979

Location	Owner/Designer/Operator	Process	Products	Capacity	Cost	Status
Baltimore County, Md.	Maryland Environmental Service; Baltimore County; Teledyne National (designer and operator)	Shredding; air classification; magnetic separation	RDF; magnetic metals; glass for secondary products; aluminum	600-1500 TPD	8.4	Shredding, air classification, magnetic separation and landfilling operational for testing; first transfer station operating
Bridgeport, Conn.	Connecticut Resources Recovery Authority; Occidental Petroleum Corp. and Combustion Equipment Assoc. (designers and operators)	Shredding; magnetic separation; air classification; froth flotation	Eco-Fuel II® (powdered fuel) for use in utility boiler; magnetic metals; non-magnetic metals; glass	1800 TPD	53^c	Construction complete; startup has begun; to be operational in 1979
Chicago, Ill. (Southwest Supplementary Fuel Processing Facility)	City of Chicago; Ralph M. Parsons Co. (designer); Consoer, Townsend & Assoc.	Shredding; air classification; magnetic separation	RDF for use by utility; magnetic metals	1000 TPD	19^d	In shakedown; began test-firing RDF; operating at 50% capacity, gradually to increase production level
Chicago, Ill. (Northwest incinerator)	City of Chicago; Metcalf & Eddy, Inc. (designer)	Waterwall combustion	Steam for Brach Candy Co.; post-incineration metals recovery	1600 TPD	23	Operational since 1971; steam delivery line under construction and expected to be on line in 1979
Columbus, Ohio	City of Columbus	Waterwall combustion; shredding; magnetic separation	Steam for electrical generation; magnetic metals	1200 TPD	118	Equipment procurement initiated; site preparation to begin in March 1979; plant to go on line by late 1981
Dade County, Fla.	Dade County; Black Clawson/Parsons & Whittemore, Inc. (designers)	Hydraposal™ (wet pulping); magnetic and other mechanical separation	Steam for utility to produce electricity; glass; aluminum; magnetic metals	3000 TPD	165^e	Contracts signed between County, P&W and Fla. Power & Light; pollution control bonds sold by state; site preparation begun; shakedown expected in 1980

Location	Participants	Process	Products	Capacity		Status
Detroit, Mich.	City of Detroit	Shredding; air classification; magnetic separation	Steam and/or electricity for use by utility; magnetic metals	3000 TPD	100	Negotiating with Combustion Engineering, Inc./ Waste Resources Corp., prior to contract signing; steam to be purchased by Detroit Edison; environmental impact statement being prepared
Duluth, Minn.	Western Lake Superior Sanitary District (operators); Consoer, Townsend & Assoc. (engineers)	Shredding; magnetic separation; air classification; secondary shredding; fluidized bed incineration of RDF and sludge	RDF; ferrous metals; steam for heating and cooling of plant and to run process equipment	400 TPD municipal solid waste; 340 TPD of 30% solids sewage sludge	19[f]	Under construction; projected startup by Apr. 1979
East Bridgewater, Mass.	City of Brockton and nearby towns; Combustion Equipment Assoc.; East Bridgewater Assoc.	Shredding; air classification; magnetic separation; other mechanical separation	Eco-Fuel II® for industrial boiler; magnetic metals	1200 TPD	10-12	Operational; fuel is being made and delivered to user
Franklin, Ohio	City of Franklin; Black Clawson Co.	Hydrasposal™/Fibreclaim™ proprietary processes using wet pulping and magnetic separation; heavy media; jigging; electrostatic precipitation	Paper fibers; magnetic metals; aluminum	150 TPD (50 TPD being processed)	3.2	Production plant operating since 1971
Hampton, Va.	City of Hampton, NASA Langley Research Center, U.S. Air Force at Langley Field, J.M. Kenith Co. (designer/builder)	Mass burning	Steam for use by NASA Langley Research Center	200 TPD	9.4	Design of large components and construction completed; delivery of boilers expected to begin in Feb. 1979
Harrisburg, Pa.	City of Harrisburg; Gannett, Fleming, Corddry and Carpenter, Inc. (designers)	Waterwall combustion; bulky waste shredding (steam driven); magnetic separation; sewage sludge burning	Steam for utility-owned district heating system and for city-owned sludge drying system; magnetic metals	720 TPD	8.3	Operational since Oct. 1972; steam main completion by Oct. 1978; sludge drying facilities completion by mid-1979

Location	Owner/Operator/Designer	Process	Products	Capacity	Cost ($M)	Status
Hempstead, N.Y.	Town of Hempstead; Hempstead Resource Recovery Corp. (Div. of Black Clawson/Parsons & Whittemore, Inc.) (owner/operator)	Hydrasposal™ (wet pulping); magnetic and mechanical separation; burning of RDF product in air-swept spout spreader stoker boilers	Electricity from utility-owned turbine generators; color-sorted glass; aluminum; magnetic metals	2000 TPD (150 TPH)	73	In shakedown
Lane County, Ore.	Lane County; Allis-Chalmers Corp.; Western Waste Corp.	Shredding; air classification; magnetic separation	RDF; magnetic metals	500 TPD	2.19	In shakedown; to be fully operational in early 1979
Madison, Wis.	City of Madison and M.L. Smith Environmental (designers); Madison Gas & Electric Co.	Shredding; magnetic separation; separation of combustibles and noncombustibles; secondary shredding air swept	RDF for use by utility; magnetic metals	400 TPD (max.) (200 TPD being processed)	2.5[h]	Under construction; startup scheduled in early 1979
Milwaukee, Wis.	City of Milwaukee; to expand to surrounding Milwaukee County areas; Americology Div. of American Can Co. (owner/operator); Bechtel, Inc. (designer)	Shredding; air classification; magnetic and other mechanical separation	RDF for use by utility; bundled paper and corrugated; magnetic metals; aluminum	1600 TPD	18	In shakedown, partially operational; test-firing RDF
Monroe County, N.Y.	Monroe County (owner); Raytheon Service Co. (designer)	Shredding; air classification; magnetic and other mechanical separation; froth flotation	RDF for use by utility; magnetic metals; nonmagnetic metals; mixed glass	2000 TPD	50.4[i]	Construction 95% complete; startup scheduled for early 1979
Nashville, Tenn.	Nashville Thermal Transfer Corp.; I.C. Thomasson & Assoc., Inc. (designer)	Thermal combustion	Steam for urban heating and cooling	500 TPD (being processed)	24.5	Operational since 1974; recently upgraded 2 boilers to 500 TPD capacity each
Newark, N.J.	City of Newark; Combustion Equipment Associates and Occidental Petroleum Corp. (designers and operators)	Shredding; air classification; magnetic separation	Eco-Fuel II® for use by utility; magnetic metals	3000 TPD (in 1000 TPD modules; to serve Newark's 700 TPD and surrounding community)	70 (for 3000 TPD) (initially 1000 TPD with a cost of $25 million including fuel user conversion)	Final contract signed 1977; some site preparation began in Dec. 1978; to be operational in mid-1980

Location	Owner/Operator	Process	Products	Capacity	Cost ($/ton)	Status
New Orleans, La.	City of New Orleans; Waste Management, Inc. (owner/operator); National Center for Resource Recovery, Inc. (designer/implementer)	Shredding; air classification; magnetic and other mechanical separation	Magnetic metals; aluminum and other non-magnetic metals; glass	700 TPD	9.1¡	Shredding/landfilling operational; recovering ferrous; aluminum, other nonferrous metals and glass in shakedown
Niagara Falls, N.Y.	Hooker Energy Corp. (Hooker Chemicals and Plastics Corp.) (owner/operator)	Shredding; magnetic separation; burning of shredded refuse	Electricity for use by company complex; magnetic metals	2200 TPD	Approximately 65	Under construction; to be operational in 1980; $20 million worth of equipment on order or on site
Pinellas County, Fla.	Pinellas County; Florida Power Corp.	Mass burning	Electricity; secondary materials recovered after burning include ferrous metals, aluminum and other non-magnetic metals	2000 TPD	80	Negotiations underway for full-service contract with UOP, Inc.; projected to begin operation by 1982
Pompano Beach, Fla.	Waste Management, Inc.; U.S. Dept. of Energy; Jacobs Engineering Co. (designer)	Shredding; air classification; magnetic and other mechanical separation; anaerobic digestion of air classified light fraction with sewage sludge	Methane gas; magnetic metals	50-100 TPD	3.6	In shakedown
San Diego County, Calif.*	San Diego County; Occidental Petroleum Corp. (designer/operator)	Shredding; air classification; magnetic and other mechanical separation; froth flotation; pyrolysis	Pyrolytic oil; magnetic and non-magnetic metals; glass	200 TPD	EPA - 4.8 San Diego County - 2 Occidental Petroleum - 8.7	Demonstration plant; operations suspended after initial testing; further possible funding and modifications being considered
Saugus, Mass.	Ten communities including Saugus and part of northern Boston; RESCO (joint venture of De Matteo Construction Co. and Wheelabrator-Frye, Inc.)	Water-wall combustion; magnetic separation	Steam for electrical generation and industrial use; magnetic metals	1200 TPD (two boilers with 600-TPD capacity each)	50	Operational since 1975
South Charleston, W. Va.	Linde Div., Union Carbide Corp.	Purox™ oxygen converter (pyrolysis); shredding	Fuel gas	200 TPD	Unknown	Demonstration plant; operated periodically since 1974

Location	Key Participants	Process	Output	Reported Capacity	Reported Capital Costs (millions of $)	Status
Tacoma, Wash.	City of Tacoma (owner/operator); Boeing Engineering (designer)	Shredding; air classification; magnetic separation	RDF; magnetic metals; steam	500 TPD	2.5[k]	In shakedown; full operation in early 1979
Wilmington, Del.*	Delaware Solid Waste Authority; EPA; Raytheon Service Co.	Shredding; air classification; magnetic and other mechanical separation; froth flotation; aerobic digestion	Ferrous metals; non-ferrous metals; glass; RDF; humus	1000 TPD municipal solid waste coprocessed with 350 TPD of 20% solids digested sewage sludge	51[l] 9 from EPA, OSW; 16 from EPA, Water Prog.; 6 from State matching grants; remainder from the Authority through sale of revenue bonds	Contract signed in Aug. 1978 with Raytheon Service Co.; design of facility is underway; groundbreaking expected in Sept. 1979

211

The following localities are either operating or constructing small modular combustion units to produce steam from mass combustion of municipal solid waste:

Operating:
Blytheville, Ark. (50 TPD)
Groveton, N.H. (30 TPD)
Siloam Springs, Ark. (19 TPD)
North Little Rock, Ark. (100 TPD)

In design:
Auburn, Maine (150 TPD)

In shakedown:
Crossville, Tenn. (60 TPD)
Salem, Va. (100 TPD)

Under construction:
Lewisburg, Tenn. (60 TPD)

In addition to the systems listed above, projects are underway to recover methane-containing gas mixtures from sanitary landfills which can be purified to pipe line quality. They are:

Azusa, Calif. — Azusa Land Reclamation Co., a wholly-owned subsidiary of the Southwestern Portland Cement Co. — Began operations in April 1978

Mountain View, Calif.* — City of Mountain View; EPA; Pacific Gas & Electric Co. — In shakedown

Palos Verdes, Calif. — Los Angeles County Sanitation District; Reserve Fuels, Inc. (joint venture of Reserve Oil & Gas Co. and NRG, Inc.) — Operational

Staten Island, N.Y. — (Fresh Kills Landfill) — New York City Resource Recovery Task Force; Brooklyn Union Gas Co., Inc.; Leonard S. Wegman, Inc.; New York State Energy Research and Development Authority — Plan to enter demonstration phase of project; preliminary testing of gas has been completed

Cost information as reported:

a Construction (including $5 million for extensions to existing steam distribution system) $31 million; engineering and construction supervision $1.5 million; interest during construction $5.5 million; contingency, start-up and land costs $1.5 million; fees, underwriting and issuance costs $2.0 million; debt service reserve fund requirement $4.5 million.

b Construction and engineering $5.6 million; land $98,000; miscellaneous equipment $165,000; plant start-up in Fall 1975 $322,000.

c Total revenues (including bond, proceeds and investment income) $54,386,040. Total expenditures: $53,386,040, consisting of the following: project development $3,026,458; bond issue expenses $1,391,413; construction $39,549,771; special capital reserve $5,022,588; debt service $5,395,810 (including main facility and six transfer stations).

d Includes design and construction. Funding through G.O. bonds.

e Land acquisition, site preparation and recovery plant $145 million; electrical power generating facility $20 million.

f Including incineration.

g Cost of Phase II of the project including construction of the resource recovery facility alone and in-plant equipment. Built in conjunction with Phase I which includes central receiving, transfer station and transfer equipment which cost approximately $2.2 million.

h For the processing plant

i Total funding authorized by county legislature; $50.4 million, including an $18.5 million grant-in-aid from New York State, D.E.C. funding under the Environmental Quality Bond Act. Includes $28.4 million for construction of the resource recovery facility. Construction of Russell Station RDF handling facility is estimated at $8 million. Balance of funds will be spent for engineering, startup, mobile equipment, etc.

j Includes Reduction Module (including landfill) $5.8 million and Recovery Module $3.3 million.

k Not including shredder which was already on-site.

l Total project costs – $51 million, including $20 million for sludge module.

*Partially funded by the U.S. Environmental Protection Agency (EPA)

Source: National Center for Resource Recovery (1979).

THE SILVICULTURAL ENERGY FARM
IN PERSPECTIVE

Jean-Francois Henry

Warrenton
Virginia

I. INTRODUCTION 216
II. HIGH DENSITY—SHORT ROTATION APPROACH TO
TREE FARMING 217
   A. Intensive Management 218
   B. Biomass Production 221
   C. Prospects and Needs 226
III. ECONOMICS OF SILVICULTURAL BIOMASS FARMS 227
   A. Conceptual Model of the Energy Farm 227
   B. Biomass Production Costs 230
IV. ENERGY BALANCE FOR BIOMASS PRODUCTION 238
V. ENVIRONMENTAL AND SOCIAL ASPECTS OF BIOMASS
ENERGY FARMS 241
VI. LAND RESOURCES: THE KEY TO ENERGY FARMING . . . 242
   A. Land Availability 243
   B. Land Utilization 244
   C. Energy Farms and the Future Wood
      Fuel/Fiber Supply 245
VII. CONCLUSIONS 246

To the memory of David J. Salo

I. INTRODUCTION

Biomass in its various forms is an attractive alternative source of energy. Through photosynthesis, biomass collects and stores low intensity solar energy which can then be harvested at will and released through direct combustion, thermochemical or biochemical conversion. Plant matter fuels have numerous advantages: they are renewable, they contain little sulfur and they have high thermodynamic availability. Furthermore, the combustion of biomass fuels does not modify the carbon dioxide or thermal balance of the atmosphere.

Because of its availability in many parts of the country and because of its demonstrated usefulness as an industrial fuel (over one quadrillion Btu of wood residues and wood processing residues are used annually by the forest products industries for process energy production), wood has received particular attention as a potential alternative fuel. It has been estimated that in the near term about 9 quadrillion Btu (quads) of wood in the form of mill and forest residues, surplus growth, annual mortality and noncommercial timber are potentially available annually for fuel use (Salo and Henry, 1979). About one-third of this potential resource could probably be collected at present at prices competitive with those of fossil fuels.

Wood therefore could make a significant contribution to reduce our dependence on fossil fuels. In the long term, around the year 2000, the projected increase in demand for fiber and the trend toward increased use of wood residues for the manufacture of products will reduce the amount of wood available for fuel. Sources of wood other than those resulting from forest products activities will therefore have to be found if wood is to become and stay a significant energy resource. The silvicultural energy farm is one of the possible sources of wood fuel in the long term (Jamison, 1977; Salo and Henry, 1979).

The concept of silvicultural energy farming originates probably in the early experiments on new silvicultural methods referred to as intensive culture (Larson and Gordon, 1979; Steinbeck and May, 1971; Ribe, 1974; Bowersox and Ward, 1976a; Saucier et al., 1972; Schreiner, 1970). These studies were concerned with increasing fiber yield to meet the projected demand for this commodity. The high yields achieved in intensively managed experimental plots encouraged the development and lent credibility to the idea of producing wood exclusively for its fuel and/or feedstock value on specialized energy farms (Szego and Kemp, 1973; Evans, 1974; Alich and Inman, 1974; Brown, 1976).

In the concept of silvicultural energy farming, agronomic techniques are used to optimize the yield of fast growing species harvested at short intervals. As it is envisioned at present, the silvicultural biomass farm involves dense planting (several thousand seedlings per acre) of fast-growing hardwood species on carefully prepared land. The farm is cultivated, fertilized and irrigated if needed using machinery similar to that used in agricultural crop production. The above ground biomass material is harvested after a few years (2 to 10 year rotations have been considered) by equipment similar to silage corn harvesters, leaving the stump in the ground. Harvesting generates wood chips which are then transported to a point of use ideally located in the center of the energy farm area. A second biomass production cycle then originates through coppice growth from the stump. Thus several harvests can be obtained from a single planting until degradation of the stump material requires a new planting.

Silvicultural energy farming involves three aspects of forest management which depart from standard forestry practice: high planting densities, short growth periods, and multiple harvests from each planting. These departures from conventional forest management introduce new issues which are discussed below.

II. HIGH DENSITY—SHORT ROTATION APPROACH TO TREE FARMING

Because of its reliance on high planting densities, short rotations and multiple harvests from a single planting, silvicultural energy farming faces management problems not generally encountered in conventional forestry. The impacts of some farm management parameters (e.g., the rate of fertilization, and the level of cultivation on the yield and cost of the fuel product) become much more pronounced under energy farming conditions than they would in conventional forestry. For instance, the nutrient balance which would establish itself naturally in long rotation situations (40-60 years) could be drastically affected when harvesting of highly productive plantings at short intervals (4-10 years) is practiced.

A number of factors such as land preparation, soil-species relationships, intensity of management of the farm and harvesting have been identified as critical for successful energy farming. These factors are all interrelated and are discussed in Section A below. Other factors such as planting density or spacing of the trees and duration of the

A. Intensive Management

Multiple harvests from a single planting are generally not feasible from evergreen farms. The following discussion is therefore limited to deciduous species.

1. Site Preparation.

Among the factors which strongly influence survival and development of the seedlings is site preparation. Site preparation, which may include clearing of existing vegetation, raking, burning, plowing and disking, is essential for the establishment of the farm (Schultz, 1969; Schultz, 1972; DeBell and Harms, 1976; Burns and Hebb, 1972; Geyer, 1974; Hunt, 1975). The degree of site preparation required will vary from site to site.

2. Planting Stock.

Planting stock adapted to the soil and climate conditions is also a prerequisite for the establishment of a highly productive farm (Briscoe, 1973; Briscoe, 1969; Randall and Mohn, 1969; Benson, 1972; Webb et al., 1973; McAlpine, 1963, Bowersox, 1973; Mohn et al., 1970).

One of the features of short rotation tree farming is that selection and breeding to improve properties of the crop can be used effectively. The use of clones allows for mass production of improved genotypes (Zavitkovski, 1978; Anderson and Zsuffa, 1975; Crist and Dawson, 1975). The use of well-tested clones and hybrids can also substantially reduce the danger of disease propagation resulting in the complete loss of a crop (Schiffer, 1976; Bowersox and Merrill, 1976; Heiligmann, 1975). The size of the planting stock—seedlings or cuttings—is important for the survival and development of the stock (Ike, 1962; McKnight, 1970; Heilman et al., 1972). Machine planting can generally be used with seedlings or cuttings of hardwoods considered as candidates for energy farming. Horizontal planting of cuttings of sycamore and green ash in shallow furrows has been tested with good results (McAlpine et al., 1972). This method of planting once perfected could significantly reduce the cost of planting in an energy farm.

3. Weed Control.

Weed control during early growth significantly improves plant survival and development and is a necessity to achieve high yields. Experience has shown that among the mechanical weed control methods tested, disking gives the best results (Hunt, 1974; Burns and Hebb, 1972;

Geyer, 1974; McKnight, 1970; Steinbeck et al., 1972; Silen, 1974). Chemical weed control has been used successfully, but caution must be exercised in the choice of the chemical agents as their use could interfere with the growth of the desired species (Burns and Hebb, 1972; Geyer, 1974; McKnight, 1970; Steinbeck, et al., 1972; DeBell, et al., 1977).

Other approaches to weed control include the use of a leguminous cover crop during early growth to inhibit weed propagation and supply some of the nitrogen required by the seedlings or cuttings. The weed control program required for successful energy farming will depend on many site specific factors such as previous use of the land, species, planting stock, climate, and others. Much experimental data must still be collected before optimum weed control programs can be identified. According to Dawson (1979), weed control may be one of the most critical features of energy farming.

4. *Irrigation.* Lack of moisture can have disastrous effects on seedling survival (Morain, 1978). It has therefore been suggested that irrigation be applied at least during the first few years of growth of a plantation even in areas where sufficient precipitation (25 inches or more) occurs during the growing season (Inman et al., 1977). It has been shown that yields of forest crops can be increased by reducing water stress during the growing season through irrigation (Broadfoot, 1964; Einspahr et al., 1972; Howe, 1968). The magnitude of the increase in yield resulting from irrigation, however, will be influenced by various site specific factors such as nutrient availability, soil quality, and length of the growing season. The potential benefits of irrigation on the overall economics of a tree farm must therefore be evaluated on a case by case basis (Mace and Gregersen, 1975).

5. *Fertilization.* The increased intensity of biomass removal and the shortening of the time interval between harvests inherent to the energy farm concept could result in serious depletion of the soil nutrient reserves and ultimately in degradation of the land. In conventional forestry practice, large stems are removed every 30 to 100 years. During the periods between harvests, branches, twigs, and leaves accumulate on the ground and at harvest times most of the noncommercial material, tops, branches, broken sections are abandoned at the logging site. The slow decay of these residues maintains the amount of soil organic matter and recycles large proportions of the nutrients absorbed by the trees during their growth cycle. Periodic harvest of

timber at long time intervals therefore creates little perturbation to soil nutrient balance.

In contrast, energy farming removes more biomass per unit area than conventional forest operations over a comparable period of time and leaves relatively less residues for nutrient recycling (Herrick and Brown, 1967; Doyle et al., 1973). The shortened intervals between harvests used in energy farming (probably 5 to 10 years) result in much shorter periods of recycling of nutrients to the soil before a new demand is made on the nutrient reserves of the soil by the next crop. The problem is compounded by the fact that young wood contains more nitrogen, phosphorous and potassium per unit biomass weight than timber age wood. The average potassium content per dry pound of 6 year old hardwoods, for instance, may be about three times as high as that of 50 year old trees (Ribe, 1974). An analysis of the nutrient budget of an aspen-hardwood stand harvested at age 30 by whole tree chipping indicates that the available soil nutrient reserves plus the nutrient inputs through mineralization, weathering and precipitation over a 30 year period exceed the removals through harvest of the crop. In contrast, the nutrient removals resulting from three harvests of 10 year old aspen sprout crops would significantly exceed the amount of nutrients available in the soil plus the expected inputs from the decomposition of residues (Boyle, 1975; Boyle et al., 1973). Nitrogen deficiencies have indeed been demonstrated in intensive cultures of cottonwood and sycamore (Blackmon and White, 1972; White, 1973). Many instances of response of short rotation crops to fertilization have been recorded. The response may, however, vary with local site properties (Happuch, 1960; White and Hook, 1975; Kormanik et al., 1973; Randall, 1973; Steinbeck, 1971; Hunt, 1974; Hunt, 1975).

The data available at present do not appear to be sufficient either to estimate exactly the needs for nutrients of short rotation hardwood plantings or to optimize the amount of nutrients for best return on the investment. In most models of energy farms (see Section III), the amounts of nutrients removed by a harvest are estimated and equal amounts of chemical fertilizers are applied during the next rotation. Estimates of the nutrients removed through harvesting can be made by determining the weight of the components of the plants—leaves, stems, branches—and using published data quoting the mineral composition of these components in young trees (Ribe, 1974; Saucier et al., 1972; Hansen, 1976; Rodin and Bazilewich, 1967). The cost of the chemical fertilizers needed to maintain the productivity of a short rotation farm is one of the major components of the cost of production of the biomass.

Several approaches have been taken to reduce this cost. One of them consists in mixing nitrogen fixing species with other hardwoods. Annual dry matter production of two year old coppice of mixed plantings of black cottonwood and nitrogen fixing red alder has been shown to be higher than that of pure cultures of either species (DeBell and Radwan, 1978). Further work is needed but this approach could improve the economics and the energy balance of the energy farm. Another method of reducing the need for chemical fertilizer is to rely on application of sewage wastewater to supply some of the nutrients needed and part of the moisture required (Sopper, 1973). This approach has the double advantage of supplying moisture and nutrients and of providing an environmentally acceptable method of disposal of urban wastewaters. Further work, however, is needed before definite conclusions concerning the impact of wastewater application to energy farms can be reached.

6. *Harvesting and Regeneration.* One of the features of the energy farm concept is that the stands will be harvested mechanically. Repeated annual cuttings from the same stumps of sycamore for the production of clone material have been performed over a period of 8 years with no detrimental effect on sprouting vigor. Stump height or angle of cutting have little effect on sprouting of cottonwood stumps (DeBell and Alford, 1972). Harvesting however should be avoided immediately following leaf development in spring and summer when carbohydrate reserves are low and stump will not sprout satisfactorily. Also, late summer harvest will result in the production of young sprouts which will be damaged by frost (Belanger and Saucier, 1975). The concept of mechanized harvesting has been demonstrated in Georiga where a corn-silage harvester has been used to cut and chip one year old sycamore stands. Specialized harvesting equipment is presently under development and should be tested in the near future. Further research is also needed to clarify the impact of the season of cutting, method of cutting, stump age and other factors on the resprouting characteristics of the stumps.

B. *Biomass Production*

1. *Species Selection.* Desirable characteristics of candidate species for silvicultural energy farms include: rapid juvenile growth, adaptability to varying site conditions, easiness to establish and regenerate, and resistance to insect and fungal diseases. Hardwoods display most of

the desirable properties for short-rotation crops, particularly those of rapid early growth and coppicing or regeneration. Hardwoods do not exhibit good site adaptability, however, and generally require a fertile and well-drained soil plus an adequate supply of moisture during the growing season (Dickman, 1975). Conifers do not fit the short-rotation criteria as well as hardwoods and therefore appear less promising for energy farming.

A representative list of deciduous species having the desirable characteristics for energy farming and the states in which they have been shown to grow satisfactorily at high planting densities on short and repeated harvest cycles is shown in Table I. At least some yield data have been reported for the species at the sites indicated by an X on the table. The important conclusion to draw from the table is that there is at least one deciduous species that has been shown to grow well under energy farming conditions essentially everywhere in the United States where adequate (25 inches or more) natural precipitation occurs. Broadly, the eastern half of the United States and the Pacific Northwest region are candidate areas for energy farming. Other regions would require systematic irrigation and are therefore less attractive. Other species have also been tested for energy farming such as black willow (Salix nigra), river birch (Betula nigra), paper birch (Betula papyrifera), balsam poplar (Populus balsamifera), oak (quercus sp.), asian tree of heaven (Ailanthus altissima) (Howlett and Gamache, 1977; White and Hook, 1975).

2. *Biomass Yields*. Among the candidate species for energy farming, sycamore (Platanus occidentalis) and various Populus hybrids have been the most extensively tested. Table II shows typical mean annual biomass production (m.a.b.p.) for the first and coppice growths of sycamore and Populus hybrids as collected by Zavitkovski (1978). Mean annual biomass production higher than those shown in Table II have been recorded: about 8-dry ton/acre-year for four year old Populus "Tristis #1" in Minnesota (Ek and Dawson, 1976), and about 9 dry ton/acre-year for hybrid poplars in Sweden (Siren and Silvertsson, 1976). Estimates suggest that m.a.b.p. of very dense red alder plantings could reach from 10 to 20 dry tons per acre year on medium to good sites (Howlett and Gamache, 1977). On the basis of the data they collected, Ek and Dawson (1976) project that m.a.b.p. of 9 to 10 dry ton per acre-year could be reached with P. "Tristis #1" for 8 to 10 year rotations.

TABLE I. A Representative List of Deciduous Species That Show Promise for Plantation Culture (Henry et al., 1976)

	Hybrid poplars		Other species											
	NE 388, 49 and 252	Others	Aspen and hybrids	Black cottonwood	Red alder	Sycamore	Pin cherry	Plains cottonwood	Eastern cottonwood	Silver maple	European black alder	Green ash	Sweetgum	Eucalyptus
New Hampshire							x							
Wisconsin		x	x											
Minnesota			x											
North Dakota								x						
Washington		x		x	x									
Pennsylvania	x	x			x									
Ohio											x			
Indiana									x					
Illinois									x		x			
Nebraska									x					
Kansas								x	x	x				
Georgia						x			x			x		
Alabama									x		x	x	x	x
Mississippi						x							x	
Louisiana									x					
Texas									x					
Florida														x

TABLE II. Mean Annual Biomass Production of Stem and Branches for First Rotation and Coppice of Sycamore and Populus Hybrids

Age (years)	Spacing (feet)	Mean annual biomass production (dry ton/acre-year)
Sycamore: first rotation		
2	1 × 4	2.1
2	4 × 4	1.1
4	1 × 4	3.8
4	4 × 4	2.6
Sycamore: coppice		
2	1 × 4	1.6
2	4 × 4	1.3
3	1 × 4	3.3
3	4 × 4	1.4
3	1 × 3	2.6
3	3 × 6	1.2
6	1 × 4	3.8
6	4 × 6	3.2
Populus hybrids: first rotation		
2	0.75 × 0.75	4.6
2	2 × 2	1.3
3	0.75 × 0.75	5.6
3	1 × 1	4.8
3	2 × 2	2.6
4	0.75 × 0.75	6.1
4	1 × 1	6.8
4	2 × 2	4.0
2	1 × 1	4.0
2	2 × 2	2.7
2	4 × 4	1.6
4	0.4 × 2.5	4.2
4	1.3 × 2.5	3.5

(Continued)

TABLE II. (Continued)

Age (years)	Spacing (feet)	Mean annual biomass production (dry ton/acre-year)
Populus hybrids: coppice		
1	2.6 × 2.6	2.6
1	3.3 × 3.3	2.3
2	2.6 × 2.6	4.8
2	3.3 × 3.3	4.5
2	1 × 3	2.2 - 8.5
2	1 × 1	4.1
2	2 × 2	5.2
2	4 × 4	3.6
4	1 × 1	6.0
4	2 × 2	6.3
4	4 × 4	5.7

Source: Zavitkovski, 1978.

3. *Rotation Age and Spacing.* It has been recognized early that a strong correlation exists between rotation age and spacing or planting density of the hardwood crop grown on the energy farm (Steinbeck and May, 1971; Zavitkovski, 1976; Gordon, 1974; Ek and Dawson, 1976; Bowersox and Ward, 1976b; Cram, 1960; Einspahr, 1972; Dawson et al., 1976; Saucier et al., 1972; Kormanik et al., 1973; Table II). The energy farm manager has the option to select a spacing when establishing the farm. Once this choice is made, the specific rotation which is reached when the mean annual biomass production culminates must be adopted in order to maximize the yield of the farm. The choice of the optimum rotation is particularly critical for close spacings (1 or 2 ft^2 per plant) generally associated with optimum rotations of the order of 3 to 5 years because productivity decreases drastically once the optimum rotation is exceeded.

The data of Table II for first rotation *Populus* hybrids illustrates this point. For a 0.75' × 0.75' planting pattern and first growth, the increase in m.a.b.p. between age 2 and 3 is about 1 dry ton/acre-year. Between age 3 and 4, the increase in m.a.b.p. is only about 0.5 dry ton/acre-year therefore suggesting that the optimum rotation for economic use of the land is of the order of 3 to 4 years for this planting pattern. For the 2' × 2' planting pattern, the m.a.b.p. increases regularly at a rate of about 1.3 dry ton

per acre-year between ages 2 and 4 with no indication of
tapering off, therefore suggesting that rotations longer than
4 years should be adopted in this case. On the basis of first
growth data, Table II suggests that close spacing and short
rotations should be favored to achieve high yields. For
coppice stands, see populus data on Table II, the difference
in m.a.b.p. between a 2' × 2' spacing and a 4' × 4' spacing
at age 4 is only about 10% of the 2' × 2' value. This suggests that wider spacing and longer rotations may be economically more advantageous for coppice growth. Selecting the
optimum spacing and rotation is a complex problem which must
take into account the growth characteristics of first and subsequent multiple coppice growth cycles to minimize the cost
of production of biomass. On the basis of the data available
at present, Steinbeck et al. (1972) and Ek and Dawson (1976b)
tend to favor spacings of at least 4' × 4' and rotations of
4 to 10, or even 15 years. Other authors prefer closer spacings and shorter rotations (Henry et al., 1976; Musnier, 1976).
Either recommendations are, however, site specific and further data will have to be collected before any "rule of thumb"
applicable in various situations can be derived.

C. *Prospects and Needs*

There is no doubt that short-rotation intensive culture
of hardwoods can achieve sustained yields significantly
higher than those achieved by conventional forestry. The
yields demonstrated so far suggest that silvicultural energy
farming could become a viable energy resource. The data presented above does show, however, that many gaps still remain
in the data base needed to fully evaluate the potential of
the energy farming concept. The many parameters involved and
their multiple interactions must be further analyzed. This
will require a concerted research and development program
extending over a long period of time. The high yields quoted
in Section II-B can only be achieved under intensive management analogous to that practiced in agricultural crop production. While such intensive management is economically justified when high valued crops are produced—food and fiber for
products—the high cost of intensive culture may be prohibitive for the production of biomass fuel or feedstock competing
with fossil fuels. The economics of energy farming are discussed in the next section.

III. ECONOMICS OF SILVICULTURAL BIOMASS FARMS

No silvicultural energy farm has been implemented so far. To estimate biomass production costs, it is therefore first necessary to develop a conceptual model of an energy farm. Production cost can then be estimated based upon the model and input data adopted. Several conceptual models of energy farms have been proposed (Inman, et al., 1977; Inter Technology/Solar Corporation, 1978; Rose, 1976; Bowersox and Ward, 1976a). These models are somewhat idealized and rely on presently available data. Actual designs of energy farms if these are implemented will be different from those adopted in the models because of site specific factors not included in the models and because of technological developments achieved at the time of implementation of the farms. With all their limitations, the conceptual models are nevertheless useful to assess the economic feasibility of energy farms and to analyze the sensitivity of biomass costs to farm management parameters.

A. Conceptual Model of the Energy Farm

The major elements included in models of energy farms include land acquisition and preparation, species selection and planting, crop management, harvesting, transportation, maintenance, and supervision. These elements are briefly described below. All costs quoted are in 1978 dollars.

1. Land Acquisition and Land Preparation. Land may be acquired through direct purchase or long-term lease. The land acquired for energy farming will generally be made of a number of parcels located within a given geographic area. The plantation density or ratio of energy farm area to the geographic area in which it is contained will have a bearing on the production cost of biomass. Among other effects, a lower planting density will increase the cost of transportation of biomass from the parcel where it is harvested to the point of use assumed to be in the middle of the geographic area containing the energy farm.

As indicated in Section II-A-1, careful preparation of the land is necessary. Preparation may involve removal of existing vegetation (brush, trees), rootraking, burning, spreading the ashes, plowing, disking, herbicide application, and work roads installation. All necessary operations can be performed with existing equipment. The cost of land preparation will vary from site to site depending among others on the amount of wooded area to be cleared, the slope and other local factors. Preparation costs are estimated to range from $125 to $375 per planted acre. Site specific studies suggest that the average preparation cost is about $200 per acre (Inman et al., 1977; Salo et al., 1979a).

2. *Species Selection and Planting.* Seedlings or cuttings of species adapted to the site must be purchased at least for the original planting of the farm. The material for replanting after the last coppice harvest has been collected can be produced on the energy farm. The current price for hardwood seedlings is about $30 to $60 per thousand. Mechanized planters can be used to plant the seedlings. The cost of the planting operation for 4' × 4' spacings with a two-row planter is estimated to be about $55 per acre (Salo et al., 1979a).

3. *Crop Management.* Crop management operations include cultivation for weed control, irrigation and fertilization. Weed control can be achieved through periodic disking between the rows of seedlings. The number of cultivations required yearly to control weeds until canopy closure occurs will depend, among other factors, on the climate and previous use of the land. For instance, a weed control program including cultivation every three weeks during the first growing season and every four to five weeks during the second growing season has been recommended as a minimum for a site in South Carolina (Salo et al., 1979a). Standard agricultural equipment is available for cultivation: regular size tractors and disks or narrow tractor-cultivator. The cost of cultivation has been estimated to be about $10 per acre per operation. This cost may be higher for close spacings where equipment of lesser capacity (acre/hour) must be used. The cost of the cultivation program mentioned above would be about $120 to $150 per acre per rotation in Southern climates.

When it is included in the energy farm model, irrigation is assumed to be applied through a traveling sprinkler system. Average capital costs for this system (well, pump, main, self-propelled gun and accessories) has been estimated to be about $115 per acre. Operation and maintenance costs have been estimated to be about $48/acre watered per year (Inman et al., 1977). A recent site-specific study by Salo et al. (1976b) shows however that site conditions such as depth of the wells and irregularities of the terrain could boost the capital costs of the sprinkler system to about $1,400 per acre. The operating costs under these circumstances would also be significantly increased. The same study shows that under the conditions prevailing at the site considered, a trickle irrigation would be less costly to install and to operate. This comparison stresses the importance of taking into account site related parameters and indicates the potential limitations of a generalized model.

As indicated in Section II-A-5, the amounts of fertilizer required by the crop are estimated on the basis of nutrient composition of the crop harvested. Typical average amounts of fertilizer needed are: about 12 pounds nitrogen (element) per acre-year per dry ton of biomass produced per acre-year, about 2 pounds P_2O_5 per acre-year per dry ton of biomass produced per acre-year and about 13 pounds K_2O per acre-year per dry ton of biomass produced per acre-year. These average amounts of nutrients assume a 60% recovery factor. Fertilizers can be applied by ground spreading, by spraying through the irrigation system or by broadcasting by aircraft. The cost of fertilization will be influenced by the method of application chosen and by the cost of fertilizers delivered at the site. Typical costs of fertilization programs including lime application once per rotation have been estimated to range between $50 and $80 per acre-year for an annual productivity of 8 dry ton per acre-year (Inman et al., 1977).

4. *Harvesting.* Following the energy farm concept, harvesting will be performed by self-propelled row harvester which will cut the stems, chip the biomass and blow the chips in a wagon following the harvester. Such a harvester could be very similar to presently used silage corn harvesters. Several tree harvesters are presently being designed under Department of Energy contracts. For lack of actual performance data, most models assume that tree harvesters will perform in a fashion similar to whole tree chippers presently used by the forest industries. To insure good regeneration of the farm (coppice crop), harvesting is limited to the dormant season in most models. The capital and operation costs for harvesting are one of the major elements of uncertainty in the estimates of biomass production costs.

5. *Transportation.* Wagons filled with chips by the harvester are emptied at field storage areas from which the chips are transported by trucks to the conversion facility. The cost of transportation will be influenced by the distribution of the planted parcels within the geographic area containing the energy farm. It is generally assumed that green chips are transported although several options of chip treatment before transportation are investigated (baling, field drying, pelletization, etc.).

6. *Maintenance and Management*. Energy farms are generally assumed to be self contained commercial operations including maintenance crews, supporting crews (fuel supplies, fertilizer supplies, etc.) as well as management and supervision personnel. The personnel required will depend among other factors on the production capacity of the energy farm and on the schedule of operations imposed by local climatic factors.

B. *Biomass Production Costs*

1. *Estimated Production Costs*. The technical data such as yields, capital and operating costs and cost of supplies for each of the components included in the conceptual designs of energy farms (see Section A above) are used as input to evaluate the economic potential of energy farms. Variable financial parameters generally include debt-equity ratio, interest rate, income tax rate and price escalation rate. Some estimated biomass production costs are shown in Table III. The production costs for Illinois and California estimated by Inman et al. (1977) are higher than other costs because it is assumed that the energy farm is planted on high priced farm land. This situation is unlikely to occur.

The very low production costs estimated by Bowersox and Ward (1976) for a Pennsylvania site are probably due to the fact that land preparation and harvesting costs for corn silage production were assumed and no fertilization or irrigation costs were included in the estimates. On the basis of the analysis of Inman et al. (1977) including irrigation and fertilization to the Pennsylvania site management would add about $11 per dry ton to the production costs at that site and would bring the cost estimates at the same level as the other costs shown in the table. With the exception of the very high and very low costs discussed above, the table shows that the costs estimated by Inter Technology (ITC) (1978) are generally lower than those estimated by Inman et al. (1977) for comparable sites. One reason for this discrepancy may be that the model used by Inman et al. (1977) assumes irrigation of the farms during the first three years of each six year rotation. If the costs of irrigation estimated by Inman et al (1977) are added to the production costs estimated by ITC for sites having comparable productivities, the two sets of estimated costs become very similar. For instance, adding $6.90 per dry ton for irrigation to the Inter Technology (ITC) (1978), estimate for the Michigan site having a productivity comparable to that of the Wisconsin site evaluated by Inman et al. (1977) brings ITC's estimate to $32.36 per dry ton or about $1.90 per million BTU as compared

TABLE III. Sample of Estimated Biomass Production Costs

Site	Productivity DT/a-y[a]	Biomass cost $/DT[b]	$/MM Btu[c]	Sources[d]
Wisconsin	5	34.72	2.04	1
Minnesota	9.1	18.48	1.09	2
Michigan	6.3	25.44	1.50	2
Michigan	5.4	31.14-36.79	1.83-2.16	3
New England	5	36.95	2.17	1
New York	6.4	20.77	1.22	2
New York	6.6	22.01	1.29	2
Pennsylvania	5	9.4-17.80	0.55-1.05	4
Missouri	7	29.08	1.71	1
Illinois	8	48.08	2.83	1
Kentucky	4.1	27.77	1.63	2
Kansas	7.5	21.69	1.28	2
Virginia	7.1	21.24	1.25	2
Tennessee	4.4	27.60	1.62	2
Louisiana	12	23.44	1.38	1
Louisiana	7.6	28.14	1.66	2
Louisiana	7.0	28.14	1.66	2
Georgia	8	26.65	1.57	1
Mississippi	12	24.39	1.43	1
Mississippi	5.6	21.46	1.26	2
South Carolina	6.5	18.76	1.10	2
Florida	12	24.17	1.42	1
Florida	7.1	19.61	1.15	2
Florida	8.0	22.76	1.34	2

(Continued)

TABLE III. (Continued)

Site	Productivity DT/a-y[a]	Biomass cost $/DT[b]	$/MM Btu[c]	Sources[d]
California	13	38.84	2.28	1
California	6.3	25.44	1.50	5
California	4.8	29.23	1.72	5
Washington	10	27.76	1.63	1
Texas	7.4	23.63	1.39	2

[a] Dry ton per acre-year.
[b] 1978 dollars per dry ton.
[c] 1978 dollar per million Btu, 8500 Btu per dry pound.
[d] Sources: (1) Inman et al., 1977; (2) Inter Technology/Solar, 1978, (3) Rose, 1976; (4) Bowersox and Ward, 1976; (5) Henry et al., 1977.

to $2.04 per million Btu as estimated by Inman et al. (1977). Similarly, adding $2.90 for irrigation to the ITC estimate for the Texas site brings the estimated production cost at that site very close to the cost estimated by Inman et al. (1977) for the Georgia site having comparable productivity. For comparable site conditions—productivity and crop management—the estimates of biomass production costs are therefore generally consistent despite the slight difference in financial assumptions and methods of estimation of the production costs. The data of the table suggest that production costs ranging from about $19 to about $28 per dry ton ($1.10 to $1.65 per million Btu) may be expected in the southeastern part of the country, production costs ranging from about $20 to about $35 per dry ton ($1.18 to $1.00 per million Btu) may be expected in the northeastern and northcentral parts of the country and production costs ranging from about $25 to about $30 per dry ton ($1.47 to $1.76 per million Btu) may be expected on the west coast. The range of costs quoted reflect site specific assumptions used in the models (productivity, land cost, level of crop management). It should be noted that the estimated production costs quoted above are comparable or slightly higher than the costs at which logging residues can be collected and delivered (Forest Service, 1978; Salo and Henry,

1979). Comparable biomass production costs have also been quoted for Canada based farms (InterGroup, 1978).

2. *Costs Breakdown.* An analysis of the production costs indicates that the operating and maintenance costs account for about 65 to 75% of the total production costs. A cost breakdown by operation as performed by Inman et al. (1977) is shown in Table IV. The highest costs are related to fertilization and irrigation (about 40% of the total) for the site considered. Other crop management costs—road maintenance, planting, weed control, harvesting, transportation—amount to about 22% of the total production costs. A cost breakdown performed by Inter Technology (1978) differs somewhat from that of Table IV in terms of the grouping of cost components. Inter Technology's analysis identifies fertilization and labor as major cost components of the production costs (about 32% and 18% of the total costs respectively). Both studies therefore suggest that improvements in the areas of crop management and fertilization in particular could have a significant impact on production costs. Inter Technology's analysis however indicates that land rental is a much higher cost component than is estimated by Inman et al. (1977). This discrepancy results probably from the combined effect of several interrelated factors such as productivity and rotation period (both of which are higher in Inman et al., 1977 case) and from the higher value of land adopted by Inter Technology (1978).

3. *Sensitivity Analysis.* The date of Table III suggest that for comparable conditions, the biomass production cost is very sensitive to productivity. Variations in productivity can result from changes in several operational parameters singly or in combinations. Cost sensitivity analyses have been performed to elucidate the impact of various parameters (Henry et al., 1976; Inman et al., 1977; Inter Technology, 1978). The results of these analyses are summarized in Table V and discussed below.

a. *Sensitivity to productivity.* As shown in Table V, all other conditions being equal, increasing productivity results in a decrease in production costs. The relative change in cost will vary according to the base case used in the analysis and can be very significant in the case of sites of low original productivity.

TABLE IV. Production Cost Breakdown by Cost Category[a]

Cost category	Production costs	
	1978 $/dry ton	% of total
Planning, supervision	1.99	8.46
Land lease	1.21	5.14
Land preparation	0.79	3.36
Roads	0.08	0.34
Planting	0.61	2.59
Irrigation	2.89	12.28
Fertilization	6.24	26.52
Weed control	0.11	0.47
Harvesting	1.06	4.50
Chips handling and transportation	3.19	13.56
Interest	0.79	3.36
Taxes	2.40	10.20
Return to investor	2.26	9.60
Salvage value	(0.09)	(0.38)
Total	23.53	100.00

[a] Data for a site located in Louisiana.

Source: Inman et al., 1977.

TABLE V. Cost Sensitivity Analyses

Parameter	Relative cost[a] variation	Sources[d]
Productivity	-0.25 to -0.72	1,2,3,4
Irrigation cost	0.1 to 0.2	1,4
Irrigation and productivity combined	could be cost effective[b]	1
Fertilization cost	0.16 to 0.20	1
Fertilization and productivity combined	could be cost effective[c]	1
Weed control program	0.02 to 0.05	4
Rotation length and spacing combined	-0.1 to +0.38	5
Number of harvests per planting	-0.25	2,3
Sensitivity to farm size		
less than 25,000 acres	-0.07	3
over 60,000 to 80,000 acres	+0.25	3
Sensitivity to size of land parcels	-0.03	2

[a] Percent change in production cost divided by percent change in the parameter analyzed.

[b] Inman et al. (1977) suggest that irrigation is cost effective if productivity increases by about 28% for a 100% increase in irrigation rate.

[c] Inman et al. (1977) suggest that fertilization over the nutrient maintenance level could be cost effective if a 40% increase in fertilization results in an 8% increase in productivity.

[d] Sources: (1) Inman et al., 1977; (2) Henry et al., 1976; (3) Inter Technology, 1978; Salo et al., 1979b; InterGroup Consulting, 1978.

b. *Sensitivity to irrigation costs.* Irrigation costs have been identified as an important component of the production costs. Increased irrigation requirements with no corresponding increase in productivity will increase the production cost. The impact is most noticeable for low productivity sites.

c. *Sensitivity to irrigation and productivity combined.* Increasing the irrigation rate (and irrigation costs) to achieve a corresponding increase in productivity may be cost effective (Mace and Gregersen, 1975). Not enough data are available at present to estimate the combined effect of irrigation and productivity. Inman et al. (1977) suggest that an increase in productivity of 28% resulting from a hundred percent increase in irrigation rate could be cost effective.

d. *Sensitivity to fertilization costs.* As expected from the analysis of Table IV, increasing the cost of fertilization, all other parameters being equal, has a negative effect on the cost of production, particularly at sites of low productivity.

e. *Sensitivity to fertilization and productivity combined.* An increase in fertilization rate over the level required to maintain the nutrient balance of the site accompanied by a corresponding increase in productivity can be cost effective. The data available at present are not sufficient to fully evaluate the cost effectiveness of increased fertilization. The analysis of Inman et al. (1977) suggests that an 8% increase in productivity resulting from a 40% increase in fertilization over the nutrient replenishment requirements could be cost effective.

f. *Sensitivity to intensity of cultivation.* Weed control has already been identified as a critical element for successful energy farming (Section II-A-3). Increasing the intensity of the cultivation program to maintain a certain level of productivity will result in an increase in production cost. The impact of this parameter, however, is probably less than that of fertilization or irrigation (Salo et al., 1979b).

g. *Sensitivity to rotation length and spacing combined.* As was discussed in Section II-B-3, rotation length and spacing are closely related. In terms of energy farming the optimum rotation/spacing combination will be the one which results in the lowest production cost of biomass over the lifetime of the farm. The analysis of Inman et al. (1977) based on first crop data suggests that a 4' × 4' spacing with rotations of about 10 years result in lowest biomass production costs. Other analyses taking into account first and coppice crops (Henry et al., 1976; Inter Technology, 1978; InterGroup Consulting Economists, 1978) suggest that closer spacings (4 to 8 ft^2/plant) and shorter rotations (2 to 4 years) should be preferred. The relative cost variations shown in Table V were estimated from data by InterGroup Consulting (1978) and should be regarded only as an indication of the sensitivity of cost to rotation and spacing; actual values of relative cost variations will depend strongly on specific site conditions.

h. *Sensitivity to the number of harvests.* Increasing the total number of harvests (first and coppice) gathered from a single planting will tend to decrease production costs.

i. *Sensitivity to the size of the energy farm.* It is generally accepted that energy farms should be large scale operations to be cost effective. An analysis by Inter Technology (1978) suggests that production costs will decrease as the size of the farm increases from about 10,000 acres to about 25,000 acres. For farm sizes between about 25,000 acres and about 80,000 acres production costs remain essentially constant displaying small fluctuations in cost around an average value as new units of production are brought in line. For larger size farms, production costs increase again with size, mostly because of increased transportation costs (Inter Technology, 1978). It should, however, be stressed that the economic viability of the energy farm will be determined by the cost of the biomass in relation to the local cost of the fossil fuels with which biomass competes. Energy farms of sizes outside of the range suggested above for minimum production may be cost effective because of local conditions (20,000-acre farm in California, Henry et al., 1977; farms smaller than 10,000 acres in Canada, InterGroup, 1978).

j. *Sensitivity to the size of individual land parcels.* Large scale energy farms will generally be made of land parcels acquired through purchase of rental. The production cost will increase as the size of the parcels of land making up the farm decrease because of increased logistics problems. The increase in production cost, however, is only about 3% when the average size of the parcels is reduced from about 1000 acres to 100 acres (Henry et al., 1976). Similar conclusions were derived by InterGroup (1978).

C. *Assessment*

The preceding discussing identifies some of the prospects and potential limitations of silvicultural energy farms.

In the present state of technology, wood fuel production through energy farming does not appear to have a significant cost advantage, if any, over wood fuel obtained from forest residues or byproducts.

As is the case for forest residues, the projected production costs would make wood produced on energy farms competitive with oil and gas for process steam generation in many areas of the United States (Arola, 1976). Competition with coal appears less promising although some studies indicate that small wood-fired power plants (10-50 MW) could be viable in rural areas (Heck, 1978). Also, as is the case for forest derived wood fuel, energy farm fuel is best suited for relatively small scale applications (up to 50 or 80 MW or equivalent steam capacity) located in the general vicinity of the source of wood fuel.

The analysis indicates that there is probably more flexibility in the design of energy farms than was originally thought: farms smaller than the often-mentioned 30,000 to 40,000 acre farms and made of relatively small parcels of land (100 acres and up) can be economically viable under certain conditions.

The analysis also points to the need for further base data before the energy farm concept can be fully and realistically evaluated. Most of the elements which were identified in Section II as being crucial for the development of the short-rotation, intensive culture approach to fiber production are equally crucial in terms of the economics of wood fuel production: productivity, optimized rotation/spacing consideration, irrigation, fertilization and weed control. There is no shortcut to generate this base data: several rotations (first and coppice) will be needed to elucidate the interactions between the various farm management parameters and their impact and farm design. It is therefore doubtful that commercial silvicultural energy farming will become a reality before the year 2000 or 2010.

IV. ENERGY BALANCE FOR BIOMASS PRODUCTION

The preceding sections have indicated that biomass fuel could be produced at costs which would make the end product of biomass conversion (steam, space heat, electricity in some cases) competitive with similar end products obtained from fossil fuels. As was discussed in Sections II and III, however the production of biomass fuel on energy farms requires intensive mangement, i.e., mechanized operations and the application of chemicals, both of which require inputs of fossil fuels/feedstocks.

For the energy farm concept to be attractive as a possible new national energy resource, the energy value of the biomass or of the products derived from biomass should be greater than the total external energy inputs. Solar energy is readily available and therefore is not included as an external input to the energy farm. The requirement of a positive net energy production efficiency assumes that the external energy inputs required to operate the energy farm are of the qualities equal to or greater than those of the biomass fuels or biomass derived fuels produced (Klass, 1977).

A sample energy budget for the energy farm model developed by Inman et al. (1977) is shown in Table VI. The energy required to manufacture the field equipment is not included in the table. Estimates by Alich and Inman (1974) and Blankenhorn et al. (1977) indeed have shown that the energy input for the manufacture of the equipment is a minor (1% or less) component in the annual energy input to the energy farm. Similarly, energy inputs related to installation of the energy farms are negligible on an annual basis over the lifetime of the farm. The data of Table VI shows that the major energy inputs are related to fertilization and irrigation. The manufacture of the fertilizer from fossil feedstocks and pumping of irrigation water are the two largest energy consuming items. Transportation and handling of the biomass is the next most energy consuming item. Not surprisingly, the same energy intensive operations are also among the most costly items identified in the biomass production costs.

The data of Table VI therefore reinforces the conclusions reached in Section III concerning the need for further research in the areas of irrigation, fertilization, and biomass handling. The use of nitrogen fixing cover crops for weed control could also decrease the requirements for chemical fertilizers and as a result improve the overall energy budget (Dawson, 1979). Different options of irrigation systems should be considered at each site to improve overall economic and energy efficiency. For instance, the recent analysis of an energy farm design for a South Carolina site by Salo et al. (1979b) shows that under the conditions prevailing at the site, a trickle irrigation system is not only less costly to install and operate than a sprinkler irrigation system but also increases the net energy efficiency of biomass production by about 20% over that projected for the sprinkler system.

TABLE VI. Sample Energy Balance for Wood Production on an Energy Farm

Operation	Energy consumed (10^3 Btu/year)	% of total consumption
Supervision	1.67	0.60
Field operations	0.53	0.19
Harvesting	7.87	2.83
Transportation and handling	26.10	9.40
Irrigation	112.48	40.52
Fertilizer	127.57	45.96
Fertilizer application	0.86	0.32
Total energy input	277.62	100.00
Total energy produced	4250.0[a]	
Net energy production	3972.0	
Net energy efficiency	0.935[b]	
Energy in:Energy out	1:15.3	

[a] 8500 Btu/dry pound
[b] Net energy produced ÷ total energy produced.

Source: Inman et al., 1977; Louisiana site, 250,000 dry ton per year, 12 dry ton per acre-year.

The net energy efficiency of biomass production will reflect site specific conditions and may therefore be somewhat different from that shown in Table VI. It is however generally accepted that very favorable energy budgets (90% net efficiency or better or energy in/energy out ratios exceeding 1:10) can be expected for energy farms (Inman et al., 1977; Alich and Inman, 1974; Inter Technology, 1978; Salo et al., 1979b; Blankenhorn et al., 1977). Blankenhorn et al. (1977) have compared the net energy efficiency of production of biomass on energy farms to that of collection of biomass from a naturally regenerated forest and concluded that both approaches to biomass fuel production have similar efficiencies.

V. ENVIRONMENTAL AND SOCIAL ASPECTS OF BIOMASS ENERGY FARMING

Silvicultural energy farming involves land management practices comparable to those required for the production of agricultural crops. A qualitative analysis by Vail and Henry (1977) suggests that, for comparable sites, energy farming would create less environmental problems than agriculture. Energy farming however will probably rely on the use of land with little or no agricultural value. Such land may have less favorable soil, slope, climate or other characteristics than prime agricultural land and therefore its use could lead to serious environmental problems. Areas of concern include air quality, surface and ground waters and soil and wildlife disruptions. Air quality could be affected by fuel engine emissions and particulate emissions produced during farm operations. The impact of these factors will probably be small as energy farms will most likely be installed in areas remote from population centers. Erosion, run off and contamination by fertilizers, soil nutrients, herbicides, and pesticides could have serious detrimental effects on surface waters. The land preparation phase during which existing vegetation is removed to plant the energy farms may be a critical period requiring particular attention. Ground cover establishment, contour ditching and other techniques are available to minimize some of the potential negative effects of energy farming. Implementation of these techniques however may be quite expensive and may reduce the competitive position of biomass fuels. Large scale irrigation may affect the water table temporarily during periods of high irrigation needs. Conversion of the land to energy farming will disrupt the soil as a result of clearing, increased exposure to sunlight and application of chemicals. The impact of these operations will be mostly felt during the installation of the energy farms. Wildlife will be affected through disruption of its habitat, disruption of its diet, and by noise and human activities. Energy farming could have some serious environmental consequences. These will have to be evaluated on a case-by-case basis. Most of the expected environmental issues can be resolved but the associated cost may be prohibitive in some cases.

Large scale implementation of energy farming will also have some social impacts. It has been estimated that an energy farm producing about 250,000 dry tons of wood per year would generate about 150 permanent and temporary jobs (Inman et al., 1977) or about 35,000 jobs per quadrillion Btu (quad) of energy produced. This increase in employment compares with that expected from the collection and delivery of forest residues for fuel use (Heck, 1978), i.e., about 30,000 jobs per quad of energy.

Other social impacts will result from the installation of energy farms within a rural area: increase in the market value of land, increase in tax revenue, stimulation of the local economy, potential influx of labor and increased need for services, aggregation of land parcels, uniformity of use and management of forest, pasture and cropland, increased traffic on rural roads, changes in aesthetic and recreational value of the land, and others. The obvious social benefit of the creation of new, permanent jobs in rural areas will have to be carefully weighted against the other social effects associated with energy farming. None of the potential negative social impacts associated with energy farming appear strong enough to prevent the implementation of this new technology but caution will have to be exercised to avoid backlash reactions which could slow the development of energy farming.

VI. LAND RESOURCES: THE KEY TO ENERGY FARMING

If silvicultural energy farming is to make a significant contribution to the national energy picture, large areas of land will have to be devoted exclusively to the production of biomass for fuel. As an example, about 42 million acres of land would have to be reserved for wood production to generate 5 quads of wood fuel annually on the basis of an average mean annual biomass production of 7 dry tons per acre year. This amount of land is less than 2% of the total land area of the United States and one could conclude that it can easily be acquired for biomass production. However, all land is not suitable for energy farming. A more realistic measure of the energy farm needs is to compare the area required for the generation of 5 quads of biomass fuel to that of cropland or commercial forest land. The area needed to produce 5 quads of wood fuel annually amounts to about 10% of either cropland or commercial forest land. Serious competition for adequate land could therefore occur between agriculture, forestry and energy farming in the future if the implementation of energy farms takes place during a period of increased demand for agricultural crops and/or forest products.

A. Land Availability

The land potentially available for energy farming has been estimated by several authors. The general criteria of suitability for energy farming adapted by these authors are a minimum of 25 inches of precipitation, arable land and slope equal to or less than 30% (17 degrees) (Salo et al., 1977; Inter Technology, 1975). Both analyses are based on the system of land classification developed by the Soil Conservation Service (USDA, 1967). The system characterizes the suitability of soils for various purposes on the basis of a scale ranging from I to VIII. Soils in Class I have few limitations to agricultural use, while soils in Class VIII are limited to recreational use. The primary uses of land in Classes I to IV are generally agriculture, pasture, and tree crops. The primary uses of land in Classes V and VI are forestry, range, watershed and some agriculture. Land in classes VII and VIII are only suitable for forestry, range, recreation, wildlife habitat and are too steep for energy farming. Both analyses involve a county by county estimate of the land potentially available in areas receiving sufficient precipitation.

Salo et al. (1977) include only land in Classes I to IV in their analysis. The most probable scenario considered by these authors assumes that 10% of permanent pasture, forest, range, rotation hay/pasture, hay land and open land formerly cropped in Classes I to IV could be diverted to energy farming. The total acreage potentially available for energy farming is estimated to be about 32 million acres and could produce about 4.5 quads annually. Inter Technology (1975) includes mostly cropland, pasture and range, noncommercial and grazed commercial forest in classes IV and VI in its analysis of land potentially available for energy farming. Only counties with a population density less than 300 persons per square mile and with a land value lower than the state average land value are included in the estimates. The authors conclude that between 75 and 100 million acres could be available for energy farming and that from 10 to 15 quads of wood fuel could be produced annually. This estimate assumes a mean annual biomass productivity of 9 dry ton per acre-year which is probably optimistic for the type of land considered (see Section II). Both studies show that most of the land potentially available is located in the eastern and central time zones.

Independently of the analyses discussed above, Dideriksen et al. (1977) have estimated that about 110 million acres of noncropland could be converted to crop production. Although about 40% of this land would be unusable for energy farming (low precipitation and/or steep slopes), this analysis confirms the fact that large areas of land could be available for energy farming.

B. Land Utilization

The previous analysis has shown that large areas of land could be converted from their present, often low productivity use, to energy farming or other uses such as food or fiber production. The amount of land which will actually be converted to energy farming will be determined by several factors. First the amount of land which could be shifted from its present use to food, fiber, or fuel production is fixed or even slightly decreasing (Doering, 1977). The conversion of medium productive farmland or even pasture to wood fuel production, if it is not associated with a corresponding increase in productivity elsewhere, could create food shortages in the long run. Therefore, policy decisions regarding the relative value of food, fiber, and fuel crops in terms of the overall national interest will have to be made. Second, all sites are not equally suitable for all crops. Climatic and other factors will dictate which crops result in the best utilization of the land, i.e., high productivity under environmentally acceptable conditions. A national policy attempting to convert all land potentially available to wood fuel production would be counterproductive as it would result in the underutilization and possible damaging of a limited natural resource, i.e., arable land.

Finally, independently of the policy and suitability factors discussed above, economic factors will also have an impact on the amount of land which will be available for energy farming. Land having the potential of producing profitable agricultural crops, if converted from an idle state to crop production will probably be used for agriculture because of the higher market value (and higher potential return on investment) of agricultural crops. Similarly, the conversion of crop land to energy farming is generally not attractive (Salo and Inman, 1977; Grantham, 1977). Several utilization options are open for land capable of producing a tree crop after the land is converted from an idle or low productive state to full utilization: pulp chips, sawlogs, veneer, fuel chips, or combinations of these. Pulp chips and sawlogs generally have a higher selling value (on the domestic or export market) than the projected selling value of fuel chips on the domestic market (Grantham, 1977). Short rotation energy farming, however, has the advantage of reducing the time period between investment and the first positive cash flow. Musnier (1976) has shown that short rotation (4-year rotation) production of wood chips and roundwood production for pulp (12-year rotation) are more attractive alternatives for farmers in the province of Quebec, Canada, than production of lumber, veneer, and pulpwood over longer rotations.

The future contribution of energy farms to the national energy supply will depend on the amount of land which can be spared for biomass fuel production. Long range planning of the utilization of the land resource will be required to insure that the national requirements for food, fiber and fuel are best achieved within the context of the present and future socio-economic constraints.

C. Energy Farms and the Future Wood Fuel/Fiber Supply

Increased demand for wood products could result in shortages around the year 2000 (Grantham, 1977). The trend towards increased utilization of wood fuel in the residential and industrial sector will further compound this supply problem. Silvicultural energy farms are one of the options available to increase the fiber/fuel supply in the long term. Another option is to increase the output of commercial forests. In 1970, the mean annual biomass production of commercial forests was about 38 cubic feet per acre-year of merchantable material or about 50 cubic feet of total biomass, i.e., about 0.85 dry ton of biomass per acre-year (U.S. Forest Service, 1973). More recent estimates (U.S. Forest Service, 1978) show that annual productivity has increased as a result of improved forest management but that it still is well below the average estimated potential of 80 cubic feet per acre year of merchantable material or 1.7 dry ton of total biomass per acre year. Reaching this potential productivity on commercial forests will help reduce the foreseen long term wood supply problem but will require a significant investment in the near term. The capital investment required has been estimated to be from $50 to $90 per acre with a corresponding increase in yield of 75 to 125 cubic feet of merchantable timber after $\sim$50 years (Spurr and Vaux, 1976). Energy farms require a higher capital investment, $150 per acre and over (Inter Technology, 1978) but their impact on the wood supply is felt over a much shorter period of time. Energy farms also offer some flexibility in terms of the products generated: changes in planting density and harvesting schedules can be implemented over short periods of time to respond to the demand for different products. The possibility of multiple crops such as the production of lumber and veneer and roundwood for pulp integrated with coppice growth for the production of wood chip has been discussed by Musnier (1978). Another option is that of using the cleaner and higher quality portion of the chip crop for products and the dirtier portion for fuel. Once the technology needed for dividing the chip stream into high and low value components is available, this option would be very attractive to the forest products industries (Jamison, 1977). As was mentioned earlier, the projected production cost and the net energy efficiency of production of biomass on energy farms are comparable to those for forest residues collected for fuel use.

It is, therefore, felt that once fully developed, energy farms will contribute to the long term supply of wood fuel. Energy farms can very usefully complement forest derived wood resources because of their "quick response" capacity to projected demands for wood fuels and products.

VII. CONCLUSIONS

The concept of silviculture energy farming is an attractive method for increasing wood fuel/fiber production to respond to the projected increased demand for these products in the future.

Available data suggest that the concept is viable, i.e., that short rotation, intensively managed hardwood plantations can reach the high annual sustained yields required for produce fuel at costs competitive with those of other energy resources. Much research is still needed before a full assessment of the potential of energy farming for fuel production can be made. However, the concept appears sufficiently promising to warrant a long term research and development program.

The impact of energy farming on the national energy resource will be determined by the amount of land which will be available for fuel production. Competition from food and fiber producers for the available land could reduce the potential contribution of energy farms to the nation's energy needs.

Much of the wood fuel supply will still be derived from forest residues in the future. However, energy farms are seen as a valuable complement to the forest resource: short rotations and some degree of adaptability in management enable energy farms to respond more quickly than forest stands to projected fuel/fiber demands.

REFERENCES

Alich, J. A., Jr., and Inman, R. E. (1974). "Effective Utilization of Solar Energy to Produce Clean Fuel." Stanford Research Institute (NSF/RANN/SE/GI/38723), SRI Project No. 2643, Menlo Park, California.

Anderson, H. W. and Zsuffa, L. (1975). "The Yield and Wood Quality of Hybrid Cottonwood Grown in Two Year Rotation." Ontario Min. Nat. Res., Div. Forests, Forest Res. Branch, Forest Res. Rep. 101, 35.

Belanger, R. P. and Saucier, J. R. (1975). Intensive culture of hardwoods in the south, *Iowa State J. Res.* 49 (3), 339-344.

Benson, M. K. (1972). "Aspen" Symposium Proceedings, USDA Forest Service, General Technical Report NC-1, 28.

Blackmon, B. G. and White, E. H. (1972). "Nitrogen Fertilization Increases Cottonwood Growth on Old Field Sites." USDA Forest Service Research Note SO-143.

Blakenhorn, P. R., Murphy, W. K., and Bowersox, T. W. (1977). "Energy Expended to Obtain Potentially Recoverable Energy from the Forest." Tappi Conference Papers, Forest Biology Wood Chemistry Conference, Madison, Wisconsin, June 20-22.

Bowersox, T. W. (1973). Influence of cutting size on juvenile growth and survival of hybrid poplar clone NE-388, *Tree Planters Notes* 24 (1) 10-11.

Bowersox, T. W. and Merrill, W. (1976). Stand density and height increment affect incidence of Septoria Canker in hybrid poplar, *Plant Disease Reporter* 60 (10), 836, 837.

Bowersox, T. W. and Ward, W. W. (1976a). Economic analysis of a short-rotation fiber production system for hybrid poplar, *J. For.* 74, 750-753.

Bowersox, T. W. and Ward, W. W. (1976b). Growth and yield of close-spaced, young hybrid poplars, *For. Sci.* 22 (4), 449-454.

Boyle, J. R., Phillips, J. J., and Ek, A. R. (1973). Whole tree harvesting: Nutrient budget evaluation, *J. For. 71*, 760-762.

Briscoe, C. B. (1973). "Extended Planting Seasons for Sycamore and Cottonwood." USDA Forest Service, Research Note SO-160.

Briscoe, C. B. (1969). "Establishment and Early Care of Sycamore Plantations." USDA Forest Service, Res. Pap. SO-50, South Forest Ext. Stn., New Orleans, Louisiana.

Broadfoot, W. M. (1964). Hardwoods respond to irrigation, *J. For. 62*, 579.

Brown, C. L. (1976). Forests as energy sources in the year 2000: What man can imagine, man can do, *J. For. 74*, 7-12.

Burns, R. M. and Hebb, E. A. (1972). USDA Forest Service, Agriculture Handbook No. 426.

Cram, W. A. (1960). *Forestry Chronicle,* September.

Crist, J. B. and Dawson, D. H. (1975). "Anatomy and Dry Weight Yields of Two *Populus* Clones Grown Under Intensive Culture." USDA For. Serv. Res. Pap. NC-113, Northcentral For. Ext. Stn., St. Paul, Minnesota.

Dawson, D. H. (1979). Personal communication.

Dawson, D. H., Isebrands, J. G., and Gordon, J. C. (1976). "Growth, Dry Weight Yields and Specific Gravity of Three-Year-Old *Populus* Grown Under Intensive Culture." USDA Forest Service, Research Paper NC-122, North Central Forest Experiment Station.

DeBell, D. S. and Harms, J. C. (1976). Identification of cost factors associated with intensive culture of short-rotation forest crops, *Iowa State J. Res. 50*, 295-300.

DeBell, D. S. and Radwan, M. A. (1978). "Growth and Nitrogen Relations of Cropped Black Cottonwood and Red Alder in Pure and Mixed Stands." Forestry Sciences Laboratory, Pacific Northwest Forest and Range Experiment Station, USDA Forest Service, Olympia, Washington.

DeBell, D. S., Strand, R. F., Peabody, D. V., and Heilman, P. E. (1977). "Coppice Management of Black Cottonwood in the Pacific Northwest." Note to be published.

Dickman, D. I. (1975). Plant materials appropriate for intensive culture of wood fiber in the north central region, *Iowa State J. Res.* 49 (3), 281-286.

Dideriksen, R., Hidlebaugh, A., and Schmude, K. (1977). "Potential Cropland Study." Soil Conservation Service, U.S. Department of Agriculture, Washington, DC.

Doering, O. C. III (1977). "Future Competitive Demand on Land for Biomass Production" in Symposium Papers Clean Fuels from Biomass and Wastes, Institute of Gas Technology, Orlando, Florida, January 25-28.

Einspahr, D. W. (1972). USDA Forest Service, Technical Report NC-1.

Einspahr, D. W., Benson, M. K., and Harder, M. L. (1972). "Influence of Irrigation and Fertilization on Growth and Wood Properties of Quaking Aspen" in Effect of Growth Acceleration on Properties of Wood Symp. Proc. Sec. I, Madison, Wisconsin.

Ek, A. R. and Dawson, D. H. (1976a). "Intensive Plantation Culture." USDA Forest Service, General Technical Report NC-21, North Central Forest Experiment Station.

Ek, A. R. and Dawson, D. H. (1976b). Actual and projected growth and yields of *Populus* "Tristis #1" under intensive culture, *Can. J. For. Res.* 6, 132-144.

Evans, R. S. (1974). "Energy Plantations: Should We Grow Trees for Power Plant Fuel?" Can. For. Serv., Dept. Environ., Rep. VP-X-129, Vancouver, British Columbia.

Forest Service (1978). U.S. Department of Agriculture, "Forest Residues Energy Program." North Central Forest Experiment Station, St. Paul, Minnesota. Final Report, ERDA Contract No. #-(49-26)-1045.

Forest Service (1978). U.S. Department of Agriculture, "Forest Statistics of the U.S., 1977." Review draft, Washington, DC.

Forest Service (1973). U.S. Department of Agriculture, "The Outlook for Timber in the United States," Washington, DC.

Geyer, W. A. (1974). *Tree Planter's Notes 25* (3), 2.

Gordon, J. C. (1974). "The Productive Potential of Woody Plants," Journal Paper No. J-7915, Iowa Agriculture and Home Economics Experiment Station, Ames, Iowa, Project 1872.

Grantham, J. B. (1977). "Anticipated Competition for Available Wood Fuels in the United States." Paper presented at the American Chem. Soc. Meeting.

Hansen, E. A. (1976) *In* "Intensive Plantation Culture" (Five Years Research). USDA Forest Science, General Technical Report NC-21, Northcentral Forest Experiment Station.

Happuch, C. D. (1960). "The Effect of Site Preparation on Survival and Growth of Sycamore Cuttings." Forest Service Southeastern Forest Experiment Station Research Note 140.

Heck, T. (1978). "The Wood Energy Concept, Its Applicability in Michigan." Report to the Upper Great Lakes Regional Commission, Michigan Department of Commerce.

Heiligmann, R. B. (1975). Weed control for the intensive culture of short-rotation forest crops, *Iowa State J. Res. 49* (3), 319-324.

Heilman, P. E., Peabody, D. V., DeBell, D. S., and Strand, R. F. (1972). A test of close-spaced, short-rotation culture of black cottonwood, *Canadian Journal of Forest Research 2*, 456-459.

Henry, J. F., Frazer, M. D., Scholten, W. B., and Vail, C. W. (1977). "Economics of Energy Crops on Specific Northern California Marginal Crop Lands." Paper presented at the Fuels from Biomass Symposium, California Energy Commission, Sacramento, California, August.

Henry, J. F., Frazer, M. D., and Vail, C. W. (1976). The Energy Plantation: Design, Operation and Economic Potential, *in* Thermal Uses and Properties of Carbohydrates and Lignins." Academic Press, New York.

Herwick, A. M. and Brown, C. L. (1967). "A New Concept in Cellulose Production: Silage Sycamore." Agricultural Science Review, Fourth Quarter.

Howe, J. P. (1968). Influence of irrigation on Ponderosa Pine, *Forest Products J. 18*, 84-93.

Howlett, K. and Gamache, A. (1977). "Silvicultural Biomass Farms: The Biomass Potential of Short-Rotation Farms." MITRE Corporation, Metrek Division, MTR-7347, Volume II.

Hunt, R. (1975). "Effects of Site Preparation on Planted Sweetgum, Sycamore and Loblolly Pine on Upland Sites." Third Year Measurement Report, International Paper Co., Southlands Experiment Forest, Bainbridge, Georgia.

Hunt, R. (1975). Technical Note No. 34, International Paper Co., Southland Experiment Forest, Bainbridge, Georgia.

Ike, A. F. (1962). Root collar diameter is a good measure of height growth potential of sycamore seedlings, USDA Forest Service *Tree Planter's Notes 54*, 9-11.

Inman, R. E., Salo, D. J., and McGurk, B. J. (1977). "Silvicultural Biomass Farms: Volume IV, Site-Specific Production Studies and Cost Analysis." MTR-7347, MITRE Corporation, Metrek Division.

InterGroup Consulting Economists, Ltd. (1978). "Liquid Fuels from Renewable Resources: Feasibility Study, Volume C, Forest Studies." Prepared for the Government of Canada, Interdepartmental Steering Committee on Canadian Renewable Liquid Fuels, Winnipeg, Manitoba.

Inter Technology/Solar Corporation (1978). "The Photosynthesis Energy Factory: Analysis, Synthesis and Demonstration." U.S. Department of Energy Contract No. EX-76-C-01-2548, Final Report, NTIS-HCP/T3548-01, Washington, DC.

Inter Technology/Solar Corporation (1975). "Solar SNG: The Estimated Availability of Resources for Large-Scale Production of SNG by Anaerobic Digestion of Specially Grown Plant Material." American Gas Association Project Number IU 114-1, Final Report.

Jamison, R. L. (1977). "Trees as a Renewable Energy Resource." Clean Fuels from Biomass Symposium Papers, sponsored by the Institute of Gas Technology, January 25-28, Orlando, Florida, 169-183.

Klass, D. L. (1977). "Biomass and Wastes as Energy Resources: Update" in Symposium Papers Clean Fuels from Biomass and Wastes, Institute of Gas Technology, January 25-28, Orlando, Florida, 2-28.

Kormanic, P. P., Tyre, G. L., and Belanger, R. P. (1973). "A Case History of Two Short-Rotation Coppice Plantations of Sycamore on Southern Piedmont Bottomlands," in IUFRO Biomass Studies, U. Maine, Orono, Maine, 351-360.

Larson, P. R. and Gordon, J. C. (1969). Photosynthesis and wood yields, *Agri. Sci. Rev. 7*, 7-14.

Mace, A. C., Jr., and Gregersen, H. M. (1975). Evaluation of irrigation as an intensive cultural practice for forest crops, *Iowa State J. Res. 49* (3), 305-312.

McAlpine, R. G. (1963). "A Comparison of Growth and Survival Between Sycamore Seedlings and Cuttings." USDA Forest Service Res. Note SE-9, Southeast Forest Exp. Stn., Asheville, North Carolina.

McAlpine, R. G., Hook, D. D., and Kormanik, P. O. (1972). Horizontal planting of sycamore cuttings, USDA Forest Service *Planter's Notes 23*, 5-7.

McKnight, J. S. (1970). USDA Forest Service, Southern Forest Experiment Station, Research Paper SO-60.

Mohn, C. A., Randall, W. K., and McKnight, J. S. (1970). "Fourteen Cottonwood Clones Selected for Midsouth Timber Production." USDA Forest Service Research Paper SO-62, Southern Forest Experiment Station.

Morain, M. (1978). Packaging Corporation, Fyler City, Michigan. Personal communication.

Musnier, A. (1976). "Etude Financiére et de gestion provisionelle des plantations et des fermes populicoles." Québec Ministére des Terres et des Forêts, Service de la Recheriche, No. 31.

Randall, W. K. (1973). USDA Forest Service Research Note SO-164.

Randall, W. K. and Mohn, C. A. (1969). "Clone-Site Interaction of Eastern Cottonwood." Proceedings of the Tenth Southern Conference on Forest Tree Improvement.

Ribe, J. H. (1974). Will short-rotation forestry supply future pulpwood needs? *Pulp and Paper*, December.

Rodin, L. E. and Bazilevich, N. I. (1967). "Production and Mineral Cycling in Terrestrial Vegetation. Oliver and Boyd Pub., London. (English Translation by G. E. Fogg.)

Rose, D. W. (1976). Cost of producing energy from wood in intensive culture, *J. Envir. Management 5*, 1-13.

Salo, D. J. and Henry, J. F. (1979). Wood Resources in the United States: Near Term and Long Term Prospects, in "Proceedings of the Workshop on Biomass and Technology," Electric Power Research Institute, Palo Alto, California, November, 1978.

Salo, D. J., Henry, J. F. and DeAgazio, A. W. (1979b). "Analysis and Design of a Silvicultural Biomass Farm." MITRE Corporation, Metrek Division, MTR-73W00102, McLean, Virginia.

Salo, D. J., Henry, J. F., and Inman, R. E. (1979a). "Design of a Pilot Silvicultural Biomass Farm at the Savannah River Plant." MITRE Corporation, Metrek Division, MTR-7960, McLean, Virginia.

Salo, D. J and Inman, R. E. (1977). "Intensive Silviculture as a Source of Energy—Fiber in California." M77-56, MITRE Corporation, Metrek Division, McLean, Virginia.

Salo, D. J., Inman, R. E., McGurk, B. J., and Verhoeff, J. (1977). "Silvicultural Biomass Farms: Volume III: Land Suitability and Availability." MTR 7347, MITRE Corporation, Metrek Division.

Saucier, J. R., Clark, A., and McAlpine, R. G. (1972). Above ground biomass yields of short rotation sycamore, *Wood Sci. 5*, 1-6.

Schiffer, A. L., Jr. (1976). "Poplar Plantation Density Influences Foliage Disease," in Intensive Plantation Culture, USDA For. Serv. Gen. Tech. Rep. NC-21, North Central For. Exp. Stn., St. Paul, Minnesota, 81-84.

Schreiner, E. J. (1970). "Mini-Rotation Forestry," USDA Forest Service Res. Pap., NE-174, Northeast For. Exp. Stn., Upper Darby, Pennsylvania.

Schultz, R. P. (1972). Silval Gentrica, 1-2,1.

Schultz, R. P. (1969. *Forest Farmer 28* (6), 8.

Silen, R. R. (1974). Forest Research Notes No. 35, Pacific Northwest Forest Experiment Stn., USDA, Forest Service.

Siren, G. and Sivertsson, G. (1976). "Survival and Dry Matter Production of Some High Yield Clones of Salix and Populus Selected for Forest Industry and Energy Production." Royal Cole, For., Dep. Reforestation, Res. Note 83.

Sopper, W. E. and Kardos, L. T. (1973). Vegetation Response to Irrigation with Municipal Wastewater, *in* "Recycling Treated Municipal Wastewater and Sludge Through Forest and Cropland." The University Press, Pennsylvania State University, 271.

Steinbeck (1971). *Can. J. of Botany 49* (3), 353.

Steinbeck, K. and May, J. T. (1971). Productivity of Very Young *Platanus Occidentalis* Plantings Grown at Various Spacings, *in* "Forest Biomass Studies" U. Maine Press, Orano, Maine.

Steinbeck, K., McAlpine, R. G., and May, J. T. (1972). *J. For. 70* (7), 406.

Szego, G. C. and Kemp, C. C. (1973). Energy forests and fuel plantations, *CHEMTECH*, May.

U.S. Department of Agriculture (1967). "Basic Statistics: National Inventory of Soil and Water Conservation Needs." Statistical Bulletin No. 461, Washington, DC.

Vail, C. W., and Henry, J. F. (1977). "Solid Fuels from Biomass: Some Environmental and Economic Considerations" Proceedings 12th IECEC Conference, Washington, DC.

Webb, C. D., Belanger, R. P., and McAlpine, R. F. (1973). "Family Differences in Early Growth and Wood Specific Gravity of American Sycamore (*Platnus Occidentalis, L)*," Proceedings of the Twelfth Conference of Southern Forest Tree Improvement, 213.

White, E. H. (1973). "Short Coppice Rotation Management of Sycamore for Cellulose Production in Western Kentucky." Paper presented at Green River Area Agents Field Day, Hainesville, Kentucky.

White, E. H. and Hook, D. D. (1975). Establishment and regeneration of silage plantings, *Iowa State J. Res.* 49 (3), 287-296.

Zavitkovski, J. (1978). "Biomass Farms for Energy Production: Biological Considerations," SAF/CIF Convention, Proceedings, October, St. Louis, Missouri.

Zavitkovski, J. (1976). *In* "Intensive Plantation Culture," USDA Forest Service, General Technical Report NC-21, North Central Forest Experiment Stn., 32.

Index

A

Agricultural residues, *see* Biomass resources
Air pollution
  control systems, 43, 162, 178–181
  problems, 112, 146, 162, 174–177, 241
  standards, 43, 175, 177–188

B

Biomass, chemicals from *see also* Methanol
  furfural, 111
  modified celluloses, 13, 16–17
  petrochemical substitutes, 6, 8, 13
  phenols, 11, 13, 111
Biomass, components of
  cellulose, 11, 12
  extractives, 12
  hemicelluloses, 12
  lignin, 12, 13–14
Biomass fuels, analysis of
  moisture contents, 29, 54, 59, 60, 62, 65, 66, 93, 151
  particle size distributions, 29
  proximate analysis of, 29, 163, 186
  ultimate analysis of, 29, 163, 186
Biomass growth, *see also* Fuel farm
  growth rates, 9, 36, 222, 224–225, 233–236
  management systems, 9, 10, 49–50, 216–226, 228–229
Biomass harvesting, 9, 29, 221, 229
Biomass, research needs, 20–22, 195–198, 226
Biomass resources
  agricultural crops, 8
  agricultural residues, 6, 8, 88, 92, 94, 101, 114, 161
  logging residues, 6, 35–39, 88, 92, 101, 114, 141, 216, 232
  manufacturing wood residues, 28–32, 33–35, 41, 90–91, 99, 111, 114, 119, 139, 141, 169, 216
  municipal waste, 8, 92, 99, 101, 114, 118
  noncommercial timber, 8, 28, 35–39, 118, 141
  silvicultural crops, 217, 218, 221–225
  spent pulping liquor, 6, 56
Biomass transportation, 54, 75, 76, 80, 229
Biomass utilization, institutional issues, 20, 20–22, 48–50

C

Carbon cycle, 2–4, 6
Cellulose, *see* Biomass components
Char, *see* Pyrolysis, products of
Coal, *see* Fossil fuels
Cogeneration
  economics of, 44, 57, 67–70, 72–73, 81
  technology of, 44, 67–68
  use of, 44–47, 67, 150
Combustion, 12, 57
  cocombustion, 112, 160–173, 178–181, 185–187
  Dutch oven, 40
  economics of, 32, 43, 48–49, 57–58, 62–66, 72–73, 81, 169–173
  efficiency of, 32, 59–62, 112, 152, 191
  fluidized bed, 42–43, 156, 157
  MSW incineration, 146, 148, 154–155, 164, 176, 181
  spreader stoker, 40–41, 57, 157, 161
  suspension burning, 43
Corrosion, 188–190

257

D

Dutch oven, *see* Combustion

E

Electricity generation, *see also* Cogeneration, 44, 70–73, 81, 150, 151, 157, 160, 165–167
Energetics, 5, 17
Energy demand and consumption, 2, 28, 55, 147, 216
Entropy crisis, 18
Ethanol, 4, 5, 11
Extractives, *see* Biomass, components of

F

Fluidized bed, *see* Combustion
Food, 2, 3, 8, 10, 20
Forest products industry, 9, 20, 216, 245–246
Forest products, *see* Wood products
Fossil fuels
  coal, 4, 6, 10, 17, 28, 56, 118, 134–136, 139, 160–163, 168, 172, 175, 183, 185–187
  natural gas, 2, 3, 134–136, 139, 163–164, 175
  oil, 2, 3, 5, 6, 28, 38, 55, 56, 64, 66, 76, 79–80, 110, 118, 146, 163–164, 172, 175
Fuel preparation
  advantages of, 42, 150–151
  costs of, 77
  drying, 41, 77, 98, 101, 106
  need for, 41–42
  systems for, *see also* Refuse derived fuels, 106, 166–167
Fuel farming
  economics, 10, 38, 227–238
  energy efficiency, 238–240
  land requirements, 242–245
  model, 227–230
Furfural, *see* Biomass, chemicals from

G

Gaseous fuels, *see also* Pyrolysis, products of
Gasification, 107–109
  air blown, 108, 123–124, 125, 126, 127–128, 153
  economics of, 57
  efficiency of, 48
  oxygen blown, 108–109, 121–123, 124, 125, 126, 153
  purification, 125, 129
  reactions of, 57, 129–130
Glucose, 10

H

Hemicelluloses, *see* Biomass, components of
Hogged fuel, *see* Biomass resources
Hydrolysis, 4

I

Incineration, *see* Combustion

L

Lignin, *see* Biomass, components of
Logging residues, *see* Biomass resources
Lumber, *see* Wood products

M

Materials
  energy competition, 6, 7, 54, 74–76, 77–78, 79–80, 216, 245–246
  reference materials system, 17–19
  substitution, 6, 7, 13–14, 216
  utilization, 6, 7, 216
Methanol manufacture
  economics of, 119, 130–134, 135, 138, 139, 142
  efficiency of, 118, 119, 120, 129–130, 136, 137
  processes for, 119–121, 122, 129–130
Modified celluloses, *see* Biomass chemicals
Municipal waste, *see* Biomass resources
  composition of, 150–151, 175
  see also Biomass, analysis of

N

Natural gas, *see* Fossil fuels
Noncommercial timber, *see* Biomass resources
Nutrient cycling, 219–220

O

Oil, *see* Fossil fuels
Oils, *see* Pyrolysis, products of

P

Paper, *see* Wood products, waste paper
Particleboard, *see* Wood products
Particulates, *see* Air pollution
Petrochemical industry, 4, 5, 13, 17, 119, 172

Index

Petrochemical substitutes, *see* Biomass, chemicals from
Phenols, *see* Biomass, chemicals from
Photosynthesis, 2–4, 6, 10, 15, 20, 216
Plywood, *see* Wood products
Pulp, *see* Wood products
Pyrolysis
  advantage of, 92
  efficiency of, 104–106, 108–109
  products of, 48, 88
    char, 88, 89–91, 94, 96, 103, 105, 111–112
    gases, 91, 94, 103, 105, 109
    oils, 91, 94, 96, 103, 105, 110–111, 153
  systems for, 48, 57, 89, 90–91, 101–104, 106–107, 113–114, 146, 153, 164

R

Refuse derived fuel (RDF)
  advantages of, 150–151, 181–184
  economics of, 190–195
  present use, 148
  processes for making, 151, 158–159, 166–167

S

Silvicultural crops, *see* Biomass resources
Solar energy, 2, 3, 6, 10
Spent pulping liquor, *see* Biomass resources
Spreader stoker, *see* Combustion
Suspension burning, *see* Combustion

W

Water pollution, 177, 241
Whole tree utilization, 217, 220
Wood products
  chemicals, 5, 6, 8, 13, 16–17
  lumber, 5, 6, 8, 9, 13, 31, 170, 172
  naval stores, 13, 15
  paper, 5, 8, 9, 13, 55, 170, 172
  particleboard, 9, 55, 74
  plywood, 9, 13, 55
  pulp, 14, 45, 54, 55, 74, 77